BENCHMARKING IN CONSTRUCTION

BENCHMARKING IN CONSTRUCTION

Steven McCabe

School of Property and Construction
University of Central England in Birmingham

Blackwell
Science

Editorial Offices:
Osney Mead, Oxford OX2 0EL
25 John Street, London WC1N 2BS
23 Ainslie Place, Edinburgh EH3 6AJ
350 Main Street, Malden
MA 02148 5018, USA
54 University Street, Carlton
Victoria 3053, Australia
10, rue Casimir Delavigne
75006 Paris, France

Other Editorial Offices:

Blackwell Wissenschafts-Verlag GmbH
Kurfürstendamm 57
10707 Berlin, Germany

Blackwell Science KK
MG Kodenmacho Building
7–10 Kodenmacho Nihombashi
Chuo-ku, Tokyo 104, Japan

Iowa State University Press
A Blackwell Science Company
2121 S. State Avenue
Ames, Iowa 50014-8300, USA

First published 2001

Set in 10.5/12.5pt Palatino
by DP Photosetting, Aylesbury, Bucks
Printed and bound in Great Britain by
MPG Books Ltd, Bodmin, Cornwall

The Blackwell Science logo is a trade mark of Blackwell Science Ltd, registered at the United Kingdom Trade Marks Registry

DISTRIBUTORS

Marston Book Services Ltd
PO Box 269
Abingdon
Oxon OX14 4YN
(*Orders:* Tel: 01235 465500
Fax: 01235 465555)

USA
Blackwell Science, Inc.
Commerce Place
350 Main Street
Malden, MA 02148 5018
(*Orders:* Tel: 800 759 6102
781 388 8250
Fax: 781 388 8255)

Canada
Login Brothers Book Company
324 Saulteaux Crescent
Winnipeg, Manitoba R3J 3T2
(*Orders:* Tel: 204 224-4068)

Australia
Blackwell Science Pty Ltd
54 University Street
Carlton, Victoria 3053
(*Orders:* Tel: 03 9347 0300
Fax: 03 9347 5001)

A catalogue record for this title is available from the British Library

ISBN 0-632-05564-2

Library of Congress
Cataloging-in-Publication Data
is available

For further information on
Blackwell Science, visit our website:
www.blackwell-science.com

Contents

Foreword

To paraphrase Charles Darwin, those companies that survive and thrive will not be the most intelligent, nor the strongest, but the most responsive to change.

Change requires learning from experience. Where are we going and what is our starting point? Sir John Egan posed the challenge in 1998 in his report *Rethinking Construction*, demanding that the industry should radically improve the processes through which it delivers products and services to its customers. Construction must be of higher quality, safer, quicker, cheaper, leaner, more predictable and more profitable. Benchmarking is the keystone of the great drive for improvement that has followed Sir John's plea.

That is why benchmarking plays a central role in the post-Egan range of initiatives driven by government and by the national institutions. Of equal importance is the amount of time and effort that busy people in the industry are prepared to invest in finding out more about the subject. A good barometer of enthusiasm is provided by attendance figures for the workshops of the Construction Productivity Network, the industry's main forum for sharing knowledge on how to do things better. Between 1998 and 2000, there were 30 workshops dealing with benchmarking, at which more than 1300 industry practitioners came together to share their experience. There was no obligation on these people to be there: all of them took time out from demanding front-line management roles because applying benchmarking in their organisations was, quite simply, essential. This is improvement in action.

This authoritative, comprehensive and admirably readable book is well timed. The initial enthusiasm for benchmarking in construction now needs to take root and grow. It must move from being a promising innovation to an everyday reality. All of those involved in planning, managing and executing construction projects should make a point of reading and absorbing its wealth of background and practical advice. Likewise, students of construction-related subjects and those teaching them will find *Benchmarking in Construction* essential reading.

Foreword

Learning to adapt to change is an urgent necessity for the construction industry - benchmarking is a vitally important means towards achieveing that end.

Gareth Thomas
Manager, Construction Productivity Network
January 2001

Preface

This book developed from research work for a PhD in construction which explored how 12 different construction companies were coping with the introduction of quality management. Carrying out this work provided an understanding of how quality management has been used in order to produce radical improvements in other industries: most especially Japanese producers of microelectronics and cars. The consequence of doing this, was to allow such companies simultaneously to reduce their costs and provide their customers with a product that was acknowledged to be superior to those of competitors.

I began to ask myself the question, if other industries can do this, why can't construction? My previous book *Quality Improvement Techniques in Construction* provided a detailed explanation of what tools and techniques are available to assist in producing radical transformation. One of the methods described was benchmarking. This is a management concept that has been specifically developed to enable organisations to discover how others have been able to produce radical improvement. The main objective of benchmarking is based on the belief that by learning how an exemplary organisation achieves things like superior customer satisfaction levels, efficiency or cost reduction, any other organisation should be able to improve its own capabilities. As a result of using benchmarking, any organisation has the potential to eventually become what is known as *world class* (usually demonstrated by winning an award for excellence).

This book describes why all construction organisations need to understand what benchmarking involves, how it has been used elsewhere to produce radical improvement, and what is involved in attempting to use it. In order to assist in understanding how the concept can be applied in construction organisations, a number of case studies written by practitioners using benchmarking tools and techniques have been provided.

Steven McCabe

Acknowledgements

There are a number of people I would like to thank for their assistance in writing this book; most especially Madeleine Metcalfe of Blackwell Science who encouraged me to develop the idea and continue writing even in my darkest hours (see dedication below). I am also grateful to those practitioners (in construction and beyond) who, over the last three years, have given their time to explain how they are using benchmarking to continuously improve. As these managers usually explain with great enthusiasm, using the tools and techniques has allowed them, their staff and their employees to learn how to improve every aspect of what they do. Their experiences have taught me the lesson that 'good enough' is never enough. We owe it to ourselves to always aim to be the best.

I would like to thank certain individuals who have assisted me in my quest to aspire to be the best. Professor Mohamed Zaira, Sabic Professor of Best Practice Management at the University of Bradford Management Centre, for his advice and guidance in what the theory of benchmarking actually involves. His commitment to helping others to improve deserves acknowledgement. Dr David Seymour of the Department of Civil Engineering at the University of Birmingham for his support and guidance during my traumatic years of attempting to develop my thesis in quality management and its application in construction. Finally to my wife Gráine and children James and John for continually encouraging me to be a better husband and father! I hope they forgive me for the time taken from them to complete this text.

Dedication

The writing of this book was seriously impeded by a profound personal loss. On the 19 February 2000, 29 days after having been diagnosed with lung cancer, my mother Margaret died. It is to her that I dedicate this book. She, of all people, always encouraged me to 'aim high', a philosophy that has served me well throughout my life. I know she was proud of my achievements, and so I offer this book to her memory.

CHAPTER ONE
INTRODUCTION

1.1 *What reading this book will assist you to do*

Benchmarking is a word that can be interpreted in different ways. As subsequent sections will explain, this book stresses the importance of benchmarking as a management concept that can assist any organisation (regardless of size or context of business) to achieve improvement. Over the last two decades the concept has been successfully used by many organisations, some of which are acknowledged to be what is called world class, and therefore among the best.

Crucially, the fact that I should consider it necessary to write a book such as this should strongly suggest my own belief in the importance of the subject. For a reader to have got this far should hopefully show a real interest in the subject of benchmarking: what it means; how it is used; and the potential benefits that may be obtained by any organisation that applies the technique. In essence the purpose of this book is to answer all of these questions. However, as with any book, the purpose of the introduction is to ensure that the reader is motivated to read further.

As will be obvious from the title of the book, this is not meant to be a racy thriller or a novel. The temptation with such books is to go to the final chapter to see how the story ends (a temptation that might negate reading the rest of the book): in the case of this book the reader is encouraged to do so. At the end of the book there are a number of case studies written by construction practitioners who have successfully used benchmarking tools and techniques to improve their organisations. These accounts are explicitly intended to be personal views; their authors have written them to explain how they themselves have implemented the theories associated with benchmarking (something I explain in the chapters subsequent to this one (see section 1.7)). Reading their accounts should enthuse the reader to finish the book and discover how to produce similar benefits.

Subsequent sections of this chapter describe influences for change by construction. Such influences have been developing for many years. However, in recent years, construction clients' belief that change is long overdue has become intense. As the book explains, the report *Rethinking Construction* (Construction Industry Task Force, 1998) made a number of recommendations of the sort of improvement that can – by learning from other industries (benchmarking) – be achieved.

1.2 People – the core concept of benchmarking for best practice

Before the reader goes on to subsequent chapters that describe the theoretical aspects of benchmarking, it should be explained that benchmarking is a tool which, despite sounding somewhat 'dry', is probably the most effective management tool to have emerged in recent years. It provides the method by which any manager, in any organisation – regardless of context – can attempt to ensure that day-to-day operations are carried out in such a way as to be 'better'. As I would suggest, very few of us can claim to be 'the best' in any aspect of what we do. Being world class – a term that will be defined later – is a term that tends to be frequently used in the context of sport. No matter how good an individual or team is, it is inevitable that others will emerge who attempt to be even better. In sport, youth is a vital part of the quest to be best. However in an analogy that is directly applicable to all organisations in business, what determines the most successful individuals or teams in sport is a mixture of the enthusiasm of youth, experience (usually from coaching staff) and dedication to the goal of competing against whatever opposition emerges. At the heart of the concept of benchmarking lies the essential element of needing to rely on the efforts of others. As the next section describes, for those who wish to be regarded as the 'best', this assumption applies regardless of the context.

1.3 Understanding the importance of benchmarking – a personal perspective

During the course of studying for my doctorate in which I explored the practical application of quality management in construction using Quality Assurance (QA) and Total Quality Management (TQM) by quality managers, I frequently tried to find distractions. One of these was to develop an interest in Formula One racing. As I began to realise, there was a connection between what I regularly witnessed in this sport – one where success is literally measured in thousandths of a second – and the use for improvement in any organisation. What became apparent by watching Formula One, was that the secret of success lies not only in state-of-the-art technology but in ensuring that a team has people who are consistently supported and encouraged to give their best at all times. As I recognised from my reading of what has been achieved in so-called 'excellent organisations' outside Formula One, their accounts suggested that they applied exactly the same philosophy to ensure success. What I also became convinced of was that with equal dedication to adopting such a philosophy, organisations in construction can also come to be acknowledged as the best.

Until the mid-1990s I had never been particularly interested in Formula One. The fact that Nigel Mansell – someone who had attended my own school in Birmingham – had won the world driver's title, interested me. As I will admit, I subsequently became a dedicated Damon Hill fan. This interest corresponded with reading articles that consistently tended to suggest one thing about organisations that had become world class, namely the application of techniques of continuous improvement which enabled the achievement of results consistently surpassing what had been believed possible in the past. As those who advocated the use of such techniques argue, in order to do this, it is necessary to constantly measure current performance against those organisations that have demonstrated the ability to achieve excellence. In Formula One, the 'battle' to be best is relentless: the difference between being first or second on the grid can be as little as thousandths of a second; the result of success is measured in millions of pounds; and, of course, there is the glory of winning races and championships.

As I reflected upon what I read and saw in Formula One I wondered what lessons there were for construction. In the former, there was the coupling of cutting-edge technology (components which, despite being unused in general production, are reliable in extreme conditions) with a driver who is able to use the car in a way that allows consistently high speed in a wide range of conditions. However, what I also became aware of was that even though a really good driver may be in the best car, without effective management and teamwork to support him,[1] the chances of success will be severely compromised. As the previous section implied people are the most essential feature of excellent organisations.

When one reads articles about quality and excellence there is a tendency to describe the experience of the electronic and automotive industries, and as a corollary, the fact that producers of these goods are frequently Japanese. In my opinion it is possible to summarise how certain Japanese producers have managed to achieve excellence by the following three words:

(1) *Dantotsu* which means the constant quest to be regarded as being the best
(2) *Kaizen* which means the obsession with continuous small-step incremental improvement
(3) *Zenbara* which means the constant search for best practice in terms of management or use of technology, and then attempting to improve upon such practice to produce something that is even more superior (and often less expensive).

The constant drive to be the best by the world class Japanese organisations is described as Total Quality Improvement (TQI) by Lascelles and Dale who believe that the consequence is an obsession with delighting the customer:

> TQI is concerned with the search for opportunities - opportunities for improving an organisation's ability to totally satisfy the customer ... the whole focus of TQI strategy will be on enhancing competitive advantage by enhancing the customer's perception of the company and the attractiveness of the product and service. This constant drive to enhance customer appeal through what the Japanese call *Miryokuteki Hinshitsu* (quality that fascinates) - the almost mystical idea that everything down to the tiniest detail has to be 'just so' - is integral to the concept of continuous improvement. Just like the concept of Total Quality *Miryokuteki Hinshitsu* is a vision, a paradigm, a value framework which will condition an entire organisational culture. This is the breakthrough, the stage at which an organisation finally breaks through to a new mind-set/paradigm: the autonomous and never-ending pursuit of complete customer satisfaction. (Lascefles & Dale, 1993: p. 294)

Chapter 3 will describe the reasons why and how certain Japanese producers of automotive and electronic goods have become *pre-eminent* in terms of *quality* and *value*. However, as these three words imply, there is a dedication on the part of such producers to continually strive to become the best, and this has involved them in constantly comparing what they do to others who are regarded as being better. These producers have successfully applied the technique of benchmarking.

There is, however, a temptation to think that the Japanese are ahead in everything. This is not always the case. Formula One – a sport that relies heavily upon the use of state-of-the-art automotive and electronic components – is one in which the British are commonly regarded as being the best in the world (especially in terms of technical and management expertise). The reason for British excellence in Formula One is, as Hotten explains, that three Ps are involved in success: products, processes, and people (Hotten, 1998: p. 228). The reason that a team can be successful for a certain period is, Hotten believes, the ability of the management of a team to ensure that it has the best possible resource input for each of these three Ps. As he explains, if a team is deficient in any of the three Ps it will rapidly suffer a loss of form. He argues that this results not only in an inability to win races (the only effective benchmark of success), but more critically in a loss of the essential funding that comes from sponsorship and partnership with those organisations whose sole motivation is success. The road back to being the best is far more difficult than remaining at the top.

Closer examination of motorsport, like its mass-production counterpart,[2] reveals that the technological innovation which is so crucial in the development of products that are faster, more reliable, and give the customer consistent satisfaction can only be achieved thorough the efforts of a dedicated and enthusiastic workforce. In Formula One, the most successful teams are those which are able to attract the best and the brightest engineers and mechanics, and crucially, have the managers most able to provide an environment which is conducive to achieving corporate ambitions (i.e. winning races), through harmony and dedication. The need to ensure that people are considered to be a vital part of the organisation's development is something that Japanese manufacturers learned from the quality guru Dr Deming in the immediate post World War II period (see Chapter 3). The need to consider people differently is something that motor manufacturers and component suppliers in the United Kingdom have, by being able to compare their methods and performance against the Japanese, realised is fundamental to producing quality. Ferry, in his book *The British Renaissance*, quotes the Chairman of Unipart – a company which experienced a transformation in the 1980s and 1990s – who makes precisely this point:

> He [John Neill] believed the secret of Japan's breakthrough in high-quality volume production was in the people. 'It's not

> investment in capital or information technology, it's investing in people. Empowering individuals, challenging them to work, and creating the space in which they can contribute their intelligence, experience and ideas'. (Ferry, 1993: p. 10)

Similarly, the Economic Intelligence Unit, in their report *Making quality work: lessons from Europe's leading companies*, describe how the Nissan plant in Sunderland has become one of the most efficient in the world despite serious reservations about the potential of this plant by senior managers in Japan. As the case study of Nissan Motor Manufacturing (UK) explains, what the workforce at this plant have achieved is now so highly regarded as to serve as best practice for all Nissan plants. In effect, the lessons that managers and workers in Sunderland learned (benchmarked) from their Japanese counterparts have been used to achieve levels of excellence that, at the very least, match all other plants. Indeed, as this report explains, the quality of cars produced is good enough for the Japanese market:

> Nissan [Sunderland] is the jewel in Nissan's overseas crown. Quality and productivity match Japanese levels. Colleagues from Spain and the USA come to learn. Best of all, the company exports cars back to Japan. (Economic Intelligence Unit, 1992: p. 13).

There are many other stories that can be told of British success (see, for example, *Achieving Quality Performance, Lessons from British Industry* edited by Teare *et al.*, 1994). What seems clear when consulting such texts is that in most cases the organisations being described needed to look to others to provide a basis upon which to build their success; in effect they benchmarked themselves against others. Reading such accounts provides a seminal lesson in the need for any organisation to understand that in order to achieve success it is essential to appreciate that others are also striving to achieve similar success. However, for many, the world of Formula One or automotive product may seem to be far removed from that of construction. As the next section of this chapter explains, the Japanese construction industry has been heavily influenced by what its own manufacturing counterpart has achieved with the use of improvement techniques.

Crucially, as another section of this chapter describes, there are some – most particularly those who co-authored the report *Rethinking Construction* (Construction Industry Task Force, 1998) – who argue that to continue to ignore such lessons is foolhardy in the extreme. As they believe, if construction organisations do not learn the lessons of other industries, their inability to satisfy clients will no longer result in disgruntlement and contractual dispute, something that has been consistently identified as being synonymous with the industry (see, for instance, Latham, 1994), but that potential work will in future be given to those who achieve precisely what the Japanese have shown to be possible. This threat, if nothing else, should make construction organisations alert to the need to engage in benchmarking for best practice.

1.4 Learning from the best: the Japanese construction industry

At the time of writing this chapter, the world seems a more uncertain place than ever. In my own lifetime (39 years), I have seen many changes. Being able to type into a personal computer is not something I would have envisaged in childhood. Strangely, however, there were things that did happen in my childhood that should have alerted me to the potential for change, the most evocative of which being a man walking on the moon. It was the need to help a man to walk on the moon which provided the stimulus for microelectronic chips to be developed, and also allows me to type this sentence into a computer which is far more powerful than anything NASA had available in the 1960s.

The history of how the Japanese saw the potential of using electronic microchips in everyday goods is now a standard part of industrial history (see, for example, Sako, 1993). As Chapter 3 will explain, whilst the Japanese may be credited with having had the foresight to see how microelectronic chips could revolutionise our world, what tends to be less well known is that it was two Americans who taught them how to use quality improvement methods to ensure that the goods they produced are now recognised as being the benchmark for excellence. In addition I describe the experience of how engineers from Toyota were sent to the enormously productive River Rouge plant that Henry Ford owned. Rather than being inspired to replicate the methods that they found there, these engineers were convinced that they could develop new systems of production that would allow cars to be manufactured as efficiently as Ford at higher quality, but most significantly, with less waste.

The ability of Japanese manufacturers to produce high-quality goods at affordable prices led many of those organisations which were in competition to re-examine their own methods. It was

American recognition of what Japanese manufacturers were capable of that led to what we now generically call TQM (Total Quality Management). Producers of electronic and automotive goods in America quickly realised that their traditional customers, people who had always bought patriotically in the past, were now opting to buy Japanese brands that were perceived to be much better. The likes of Ford and Chrysler, for instance, knew that they had to respond. In essence, they had, to quote the cliché , to 'Get better, or get beaten'.

As Bank describes in his book *The Essence of Total Quality Management,* at the heart of the philosophy of TQM is the presumption that the customer is 'king' (Bank, 1992: p. 1). As he argues, the response of any customer who is disappointed by the quality of what they receive will be to withdraw their business and buy elsewhere. Accordingly this has become one of the assumptions which dominates our everyday lives; we are all 'kings' and as a consequence we owe loyalty to no organisation, unless, there are good reasons to do so. The ability to attain extremely high levels of customer satisfaction is something that Japanese producers of cars and electronic goods have dedicated a great deal of effort to achieving. As a consequence, it has been shown that consumers are much more likely to remain loyal to the company from which they purchased the item. It was therefore inevitable that because of what Japanese manufacturing had achieved using quality improvement techniques, its construction industry would seek to emulate their experiences.

One excellent piece of literature that I would recommend which describes how Japanese construction is organised is a report that was carried out on behalf of the Chartered Institute of Building (CIOB), *Time for real improvement: learning from best practice in Japanese construction R&D* (1995). The message this report contains is as pertinent now as it was at the time of publication. For instance, as Herb Nahapiet – the leader of the mission that went to Japan – argues in the foreword to this report:

> With the UK construction industry fighting to survive in an increasingly competitive international marketplace, a report on research and development might seem remote and irrelevant. The reverse is the case. Without coherent, comprehensive, long-term initiatives such as those in Japan, the cycle of decline and low profitability in Britain's construction industry will continue, eventually blunting its competitive edge. (Chartered Insitute of Building, 1995: p. i)

If these words may have seemed unduly pessimistic at the time, the recent experience of construction suggests that his prediction was accurate. British construction, it seems, is more in need of radical change than ever. Therefore a reconsideration of what the CIOB report discovered is timely in attempting to understand how it is possible that British construction, by benchmarking itself against Japan, can learn to improve.

The report proposes a model which, according to Nahapiet is summarised by the acronym **TIME**:

- Technology
  The Japanese are technological master craftsmen. When they talk about improvement, it is through technological, rather than managerial change.
- Innovation
  The form of R&D which the Japanese undertake is nearer to innovation than that which is conventionally undertaken in the UK. Their research normally involves the observation of real systems with a view to continuous improvement.
- Motivation
  At all levels in Japan there is real commitment to innovation and continuous improvement.
- Empowerment
  All parties in the Japanese construction R&D processes feel themselves to be empowered to get on with the job and are properly supported by their leaders.

In describing how post World War II Japan literally rebuilt itself to become one of the most dominant forces of the industrialised world, the report explains that in order to achieve this transition, the Japanese learned (benchmarked) what was being done in the West:

> Japan emerged from World War II with millions of people homeless, its industry in ruins, and totally dependent on the outside world for the majority of its raw materials. Such was its economic success that by 1952 output had reached pre-war levels, only to double and then treble a few years later. By the middle of the 1960s it had exceeded the UK's economy, and by the end of the decade, Germany's. In 1990 it overtook the USA in per capita GNP.
>
> During this period, as throughout its history, Japan demon-

> strated its ability to absorb revolutionary changes from outside especially the West and through processes of continuous improvement to enhance the knowledge gained at no cost to, indeed largely on the basis of, its traditional values. (Chartered Institute of Building, 1995: p. 3)

In order to create the conditions that would allow this phenomenal development, it was inevitable that organisations operating in Japanese construction should dedicate themselves to continuous improvement. As the report explains, an essential element that has enabled construction to be able to respond to the demands of the industrialised clients that have been so vital to Japan's extraordinary post-war development has been a recognition of the significance of investment in research and development.

Whilst a full description of how the process of funding of research and development is beyond the scope of this book, it is worth reviewing the main findings that the CIOB mission to Japan discovered:

> The traditional top-down, linear model used to explain the processes of British R&D was not found appropriate to successful Japanese practices. Instead, the Mission team has developed a more interactive and cyclical model. Central to it are a linked series of real improvement cycles throughout a company. Relentless self-scrutiny is combined in this model with a strong outward, benchmarking focus. (Chartered Institute of Building, 1995: p. 20)

The model that was proposed is shown below in Fig. 1.1. (RICARQ is Real Improvement Cycles Against Recognisable Qualities.)

There are various examples of the way that Japanese construction has used this model to create products (buildings) that are not only more innovative than those found in the West, but are created far more efficiently. For instance, Cargill reports on how houses are built there:

> Japan has only twice the population of the UK but is building more than eight times as many new houses ... By taking a lead from manufacturing and introducing standardised components, modular building and a semi-automated construction process, Japan has speeded up production and halved labour costs. This has allowed more money to be put into better-quality materials and designs that will last longer and look better. (Cargill, 1994a: p. 33).

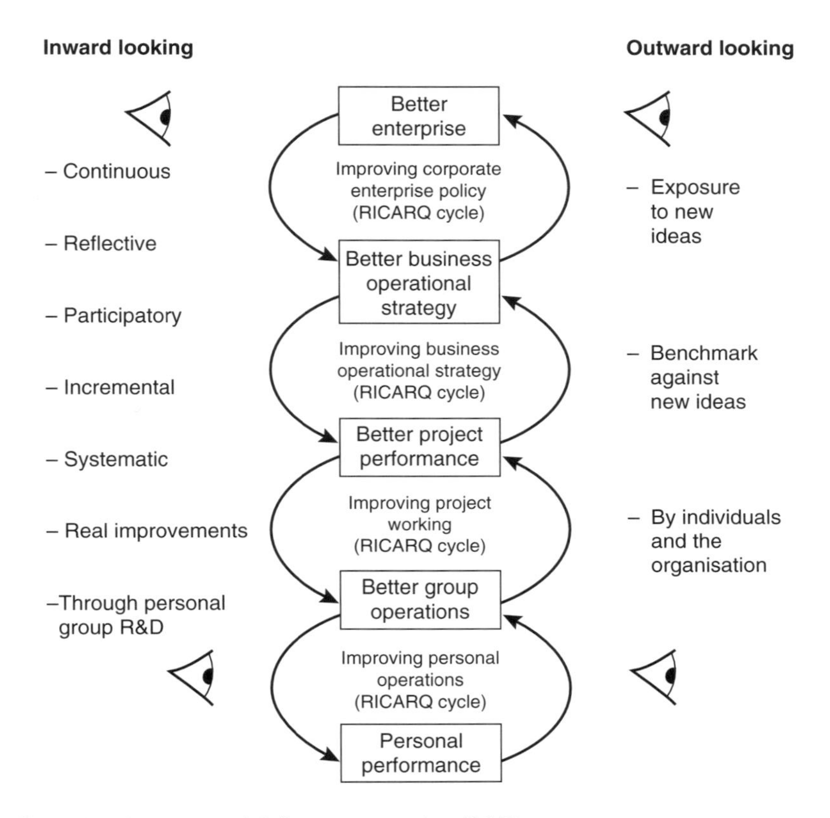

Fig. 1.1 A new model for construction R&D.

As Cargill stresses, whilst prefabricated building has a negative image in this country, the houses that the Japanese produce are 'indistinguishable from those built entirely on site' (Cargill, 1994a). Moreover, as she explains, the factory conditions that these houses are produced in allow for considerably higher levels of quality control than would be possible on site. Interestingly, a report in the *Financial Times* describes how, because of a demand by the Japanese for all things English, there is a rapidly expanded market for 'traditional' timber-framed houses (Taylor, 1996). As the report explains, these buildings, which come in kit-form, can be rapidly constructed to very high standards.[3] It is precisely this use of production methods and organisation that has enabled the fast-food chain McDonald's to construct restaurants in less than two days:

> [The modular] approach has allowed McDonald's to put up entire buildings in less than a day and a half. The modular system

> saves time and money by cutting down on design time and standardising the construction process. (Cargill, 1994b: p. 34)

Other examples exist of how the use of prefabrication, more highly trained operatives and a radically different relationship between client and suppliers (builders and subcontractors) allows for the production of buildings which have the advantage both of being extremely efficient to construct and achieving very high standards of quality. Therefore, logically, the question can be asked: 'If these examples exist, and the benefits that emanate from them are so well-known, why are the lessons not more widely applied?'. As the next section of this chapter describes, one of the most influential reports to be published on construction in recent years, makes exactly this argument.

1.5 Rethinking Construction: *a catalyst for change in British construction?*

Ball, in his seminal book *Rebuilding Construction* believes that what the construction industry produces suffers from a perception of not being very good:

> 'Technology' must partly be to blame, but far more important is the organisation of the construction process. Fragmentation of responsibility for production is a recipe for building failures, especially in a context where everyone has an incentive to work at speed and save time by cutting corners. (Ball, 1988: p. 217)

As Ball suggests, even though productivity of British construction appears to have dramatically improved in the 1980s, its achievement was at the expense of quality. As a consequence, he concludes, 'Within the current structure of the industry, it is difficult to see how the quality problem can be surmounted' (Ball, 1988). Ball was not the first to have identified these sorts of problems, nor that their cause was due to the contractual arrangements that exist to 'regulate' the relationship between client and those who supply the finished product. Any student of construction will note that there has been a succession of studies carried out into this industry in Britain since World War II (for example, Simon, 1944; Banwell, 1964; Wood, 1975). The precise effects that these reports have had in changing the British construction industry is debatable. However, the authors of these reports should be applauded for at least attempting to present alternative models of how the construction industry can be better organised. Moreover one thing that all these reports tend to agree upon is that the resources used by the construction industry can be made to perform more effectively, and that as a direct result parties could enjoy the benefits.

Two of the most recent reports into the construction industry (Latham, 1994; Construction Industry Task Force, 1998)[4] are notable for the following:

(1) The power of the assertions that they contain
(2) The fact that, subsequent to publication, the stated intention of Government is to ensure that recommendations contained within them are implemented.

Latham recommended that alternative arrangements for contractual relationships were an essential element in creating improvement in construction. By so doing, Latham asserted, it should be possible to attempt to achieve in the order of 30% improvement in productivity. The Egan Report, whilst being entirely sympathetic to Latham's recommendations, proposed that much more needed to be done in order to achieve the sort of radical improvement that many have suggested is possible in construction. As Sir John Egan asserts in the foreword: 'At its best the UK construction industry displays excellence. But there is no doubt that substantial improvements in quality and efficiency are possible' (Construction Industry Task Force, 1998: p. 5)

The Egan Report created debate because of what many saw as the very challenging recommendations that were made in the magnitude of improvement that, the authors believed (on the basis of 'experience and evidence'), were possible. The most notable of these were that:

> Our targets include annual reductions of 10% in construction cost and construction time. We also propose that defects in projects should be reduced by 20% per year. (Construction Industry Task Force, 1998: p. 7)

Dr Deming[5] was reputed frequently to pose the question to those who wanted to achieve change, 'By what means?'. This question is absolutely pertinent in considering how construction can achieve the sort of dramatic improvement that the Egan Report believes is possible. As the report suggests, if construction is to create conditions favourable to radical improvement it must make certain changes, the most notable of which are:

(1) Modernise
(2) Increase spending on training and research and development
(3) Create better relationships between contractors and clients
(4) Increase the use of standardisation and pre-assembly
(5) Apply performance tools and techniques

With respect to the last of these, there are three techniques that are specifically referred to: CALIBRE[6]; value management[7]; and benchmarking. All of these techniques are of interest in that they provide ways for managers to introduce the concept of improvement and best practice into their organisations. Whilst CALIBRE and value management are highly beneficial tools, it is the last of the three tools, benchmarking, that this books explains. According to the Egan Report, benchmarking 'is a management tool which can help construction firms to understand how their performance measures up to their competitors' and drive improvement up to 'world class' standards.[8] (Construction Industry Task Force, 1998: p. 13)

The report, which includes representatives from organisations such as Tesco, Nissan, British Steel and Cardiff Business School, argues that construction can learn from the experiences of other industries. Specifically, the report refers to manufacturing and service industries in which, the authors assert, there have been 'increases in efficiency and transformation of companies which a decade ago nobody would have believed possible' (Construction Industry Task Force, 1998: p. 14). Therefore, it is argued, if construction is to achieve the sort of radical improvement that has been produced in these industries, it must be prepared to be committed to five 'fundamentals to the process' (Construction Industry Task Force, 1998: p. 16).

These five fundamentals are (each of these quotations from the Egan Report are exhortations for change by the authors):

(1) Committed leadership

> ... we have yet to see widespread evidence of the burning commitment to raise quality and efficiency we believe is necessary. (Construction Industry Task Force, 1998)

(2) A focus on the customer

> ... the construction industry tends not to think about the customer ... [there is] little systematic research on what the end-user actually wants, nor to raise customers' aspirations

> and educate them to become more discerning. We think clients, both public sector and private sector, should be much more demanding of construction. (Construction Industry Task Force, 1998)

(3) Integrate the process and the team around the product

> ... construction typically deal[s] with the project process as a series of sequential and largely separate operations undertaken by individual designers, constructors and suppliers who have no stake in the long term success of the product and no commitment to it. Changing this culture is fundamental to increasing efficiency and quality in construction. (Construction Industry Task Force, 1998)

(4) A quality driven agenda

> [construction] must understand what clients mean by quality and break the vicious circle of poor service and low client expectations by delivering real quality. (Construction Industry Task Force, 1998)

(5) Commitment to people

> ... construction does not yet recognise that its people are its greatest asset and treat them as such. [it] cannot afford not to get the best from the people who create value for clients and profits for companies (Construction Industry Task Force, 1998)

As will be shown in subsequent chapters, the concepts that underpin TQM and continuous improvement are very similar to these five fundamentals. As those who advocate the use of such a philosophy stress, in order to succeed it is vital that there is a combination of demonstrating improvement of processes carried out and giving people a greater level of importance than hitherto. In order to demonstrate improvement, it is essential that specific measures are used. The Egan Report argues that construction must institute such measures:

> To drive dramatic performance improvement the Task Force believes that the construction industry should set itself clear measurable objectives and then give them focus by adopting qualified targets, milestones and performance indicators. (Construction Industry Task Force, 1998: p. 17)

Table 1.1 – which has been adapted – shows the assessment of 'minimum scope for improvement in the performance of UK construction' (Construction Industry Task Force, 1998).

These indicators – sometimes called *key performance indicators* – are now being used to provide targets for all organisations that operate in the construction industry. What these are and how they

Table 1.1 The Construction Task Force targets for improvement

Indicator	Percentage improvement per year	Current performance being achieved by leading clients and certain construction companies
Capital costs (i.e. all costs excluding land and finance)	Reduce by 10%	Evidence that some clients have been able to reduce costs through supply chain of up to 14% in last five years
Construction time (i.e. time from client approval to practical completion	Reduce by 10%	Some UK clients and design-and-build firms have reduced construction time on offices, roads, stores and houses by 10–15%
Predictability (i.e. number of projects completed on time and to budget)	Increase by 20%	Evidence of clients increasing predictability by 20% per year and regularly achieving 95% certainty
Defects (i.e. reduction in number on hand-over to client)	Reduce by 20%	Some clients are striving to achieve zero defects. Evidence that contractors can also do this
Accidents (i.e. as reported to HSE)	Reduce by 20%	Clients and contractors have shown that it is possible to reduce accidents by 50–60%
Productivity (i.e. increase in 'value added' per employee)	Increase by 10%	Evidence that 5% already being achieved by some contractors, and that best projects in UK and US achieve up to 15% per year
Turnover and profits (i.e. of the construction companies employed to carry out work)	Increase by 10%	Best construction companies have increased turnover and profits by 10–20% and, correspondingly, they are raising their profit margins as a proportion of turnover well above the industry average

can be used by any organisation that wishes to improve is described in Chapter Five.

In order to reinforce the message that improvement is achievable, the Egan Report provides examples of what organisations have been able to achieve using benchmarking and quality management techniques. For instance:

- Tesco, who have reduced the cost of building new retail outlets by 40% whilst at the same time decreasing the time required
- BAA being able to reduce time taken on construction by more than 30%
- The use of 'lean construction' in the USA by Neenan and Pacific Contracting to reduce cost, increase predictability and improve client satisfaction

Included in the conclusions of the Egan Report is the following statement:

> There is an urgent need for the construction industry to develop a knowledge centre through which the whole industry and all of its clients can access knowledge about good practices, innovations and the performance of companies and projects. (Construction Industry Task Force, 1998: p. 39)

As the following sections explain, a programme exists that is dedicated to propagating the aspiration of achieving best practice in construction.

1.6 The Construction Best Practice Programme

In a book dedicated to informing practitioners on how to achieve benchmarking and best practice, it is satisfying to be able to report on the work being carried out by this project. It is recommended that anyone wanting to discover more about the techniques described in subsequent chapters should contact the Construction Best Practice Programme (see section 8.4.3 of Chapter 8). Prior to doing this, it will be useful to the reader to have some knowledge about what this programme purports to do, and how it carries out its work.[9]

The emergence of the Construction Best Practice Programme in November 1988 came under the auspices of the Construction Industry Board (CIB).[10] In July 1997 the Minister for Construction at the Department of the Environment, Nick Raynsford, announced the desire of Government to work with industry through CIB to develop an initiative which would achieve the following objectives:

- Create the desire for improvement by publicising the activities of successful construction organisations
- Show the benefits that are possible through improved practice
- Provision of a first point of contact for construction organisations wishing to attempt to improve
- Identify, publicise and support the implementation of business improvement tools, techniques and advice
- Provide a means by which communication is effectively facilitated between those organisations that wish to improve and those organisations that possess the experience and knowledge of how to achieve this objective
- Disseminate research which shows the potential for improvement and benchmarking in the construction industry supply chain

In pursuance of these objectives, the Construction Best Practice Programme offers the following programme services.

Information line

There are a number of fact sheets which provide introductory material which can be obtained via telephone enquiry or directly through the website (see section 8.4.3)

IUKE–construction company visits

This collaborates with the Inside UK Enterprise Scheme which the Department of Trade and Industry operates, and seeks to provide an 'open door to understanding current best practice in some of the best construction enterprises' (quoted from Construction Best Practice Programme literature, date unknown). These visits operate in the following way:

(1) A *host* organisation provides access to outsiders in order to understand how they are applying any aspect of improvement or best practice.[11] It allows such organisations to be recognised as being a leader in the application of the initiative being demonstrated.
(2) *Visitors* derive benefit from being able to see in action the

initiative being carried out within the host organisation and, in particular, being able to talk to those directly involved about how it may be possible to attempt a similar initiative in their own organisation.

(3) By being involved in these visits, the Construction Best Practice Programme is 'able to provide a barometer of the industry's current thinking on best practice'.

Workshops

These are provided by the Construction Productivity Network, and aim to promote the sharing of knowledge and understanding of the range of tools and techniques that exist to assist in achievement of improvement.

Case studies

These relate to so-called 'levers of change' that have been adopted by the Construction Best Practice Programme. As such they are usually provided in the form of a leaflet which is intended to provide a synopsis of how a case study organisation implemented an improvement initiative, what benefits they have enjoyed and how it is possible for any other organisation to try to adopt a similar approach.

Advice signpost services

One of the descriptions that have been applied to the Construction Best Practice Programme is that it acts as a 'sort of dating agency – they help you find the right partner'. This analogy is verified by literature provided by the programme (undated and no pagination):

> This service directs enquirers to companies who can give direct advice on best practice and associated business improvement opportunities. Helpdesk staff identify the enquirer's needs and then direct them to contacts who are competent to provide the right level of advice.

Champions for change

This is a forum that exists whereby senior members of organisations engaged in the implementation of best practice share their expertise with others.

Construction Best Practice in Action

This is a tool provided on a CD-Rom basis that seeks to provide practical advice to a small organisation on how it might be possible to improve performance.

Best Practice clubs

A number of these exist throughout the country, usually with the specific purpose of enabling representatives from different organisations to exchange advice and guidance in the use of improvement tools and techniques.

As the Construction Best Practice Programme explains, its role is not to tell any organisation what it should do or how it should do it. As is usually the case with any new management concept, many consultants exist who will attempt to do this (at a cost of course). The Construction Best Practice Programme aims to provide encouragement and assistance to those who wish to apply improvement techniques to their own organisation. As the next section describes, one of the most recognised and effective ways of doing this is by the use of what are known as Key Performance Indicators.

1.6.1 The key performance indicators

Chapter 5 provides a detailed explanation of critical success factors and key performance indicators (KPIs). The former are the statements of *how* improved business practice must be achieved if an organisation is to be able to attain its mission. The latter are the *means* by which an organisation can measure the progress being made to ensure that the critical success factors are being achieved.

In these sentences the word *organisation* is used in its widest sense, and in particular, to include a society or body of members. Therefore, the British construction industry can be collectively considered as being an organisation made up of many small members, all of which exist for their own particular purposes. However, the construction industry – like any industry – is usually believed to act collectively to ensure that all of its members obtain maximum benefit. It is precisely for this reason that the Department of Transport and the Regions (DETR), through its Minister for Construction attempts to co-ordinate the efforts of all those involved in construction.[12] As previously discussed, one of the main recommendations to emerge from the authors of *Rethinking Construction* was that the industry should 'put in place a means of measuring progress towards its objectives and targets towards improvement' (Construction Industry Task Force, 1998: p. 18). Such measures would therefore be the KPIs that could be used to judge how effective construction was being in implementing the sort of improvement envisaged in *Rethinking Construction*. Indeed, as the DETR explains in its publication *KPI Report for The Minister for Construction*,[13] the importance of KPIs is that they allow the 'measurement of project and organisational performance throughout the construction industry' (DETR, 2000: p. 7). As the DETR argues, the information that such KPIs generate makes it possible for clients and 'supply chain' organisations to effectively engage in benchmarking 'towards achieving best practice' (DETR, 2000).

There are ten KPIs that are currently being used. These are now discussed in turn.

(1) *Client satisfaction – product.* This measures the satisfaction level of a client with the finished product they received and uses a ten-point scale where ten is highest.

(2) *Client satisfaction – service.* This measures the satisfaction level of a client with respect to the service they received from the consultants or main contractor they employed (as in (1), a ten-point scale is used).

(3) *Defects.* This measures the 'condition of the facility' at the time of hand-over with respect to defects and uses a ten-point scale in which:

10 = Defect-free
8 = There are 'some defects' but which have no 'significant impact' on the client
5 = There are defects, some of which have 'impact on the client'
3 = There are 'major defects' which have a 'major impact on the client'
1 = The facility is 'totally defective'

(4) *Safety.* This measures the reportable accidents per 100 000 employees. An accident is defined as being reportable by the Health and Safety Executive if it results in death, major injury or over three days sickness to employees, those who are self-employed or members of the public.

(5) *Predictability – cost.* There are two elements to cost:

(a) Design, which is defined as being 'actual cost at available for use less the estimated cost at commit to invest,[14] expressed as a percentage of the estimated cost at commit to invest'.
(b) Construction, which is defined as being 'actual costs at available for use less the estimated cost at commit to construct,[15] expressed as a percentage of the estimated cost at commit to construct'.

(6) *Predictability – time.* There are two elements to time:

(a) Design, which is defined as being 'actual duration at commit to construct less the estimated duration at commit to invest, expressed as a percentage of the estimated duration at commit to invest'.
(b) Construction, which is defined as being 'actual duration at available for use less the estimated duration at commit to construct, expressed as a percentage of the estimated duration at commit to construct'

(7) *Construction time.* This is the normalised time (a statistical method which takes account of location, function, size and inflation) to construct projects when a comparison is carried out from year to year.

(8) *Construction cost.* This is the normalised cost (see (7)) for definition of normalisation) of projects when taken in comparison from year to year.

(9) *Productivity.* This is the measure of the average value that has been added by each employee (total value is turnover less all costs subcontracted to, or supplied by, other parties).

(10) *Profitability.* This is the amount left prior to tax and interest as a percentage of sales.

According to the Construction Best Practice Programme, anyone can use these KPIs. Therefore, clients, designers, consultants, contractors and subcontractors can consider how their organisations compare to the data that will have been collected from a large number of organisations by the Construction Best Practice Programme in the preceding year. As such, the use of such comparisons enables organisations to consider the following:

- The progress made in particular areas of their business
- The potential for implementing initiatives for producing improvement
- The need to do more than simply measure KPIs

A crucial feature of the concept of benchmarking is the need to understand processes, and in so doing seek alternative ways of carrying out the day-to-day activities which are fundamental to completion of the overall corporate objectives (see Chapter 5). As this book will stress, whilst measurement is probably the most essential part of carrying out benchmarking, it is important to understand what the measures are and how effective they might be in producing improvement. It is for this reason that attention needs to be drawn to the last of the above bullet points which strongly implies that the outputs of measurement should not be considered to be the only activity. There is a danger in thinking that merely because your organisation is better than others competing in the same sector, that will be enough.

Benchmarking is a management tool that can allow any organisation to consider what it does and how it achieves it in comparison to any other organisation, regardless of the fact that it may operate in a sector which is entirely unrelated to construction. If there is a desire to improve the capability of an organisation radically the most effective way of doing this will be to compare it against the 'best in the business'. As the authors of *Rethinking Construction* suggest, a construction organisation should not simply aim to be as good as any other construction organisation, but should aim to use benchmarking as a tool to 'drive improvement up to "world class" standards' (Construction Industry Task Force, 1998: p. 13). World class is an expression that has been used in this chapter to imply that an organisation is accepted as being the 'best in the business'. Organisations that achieve the accolade of being world class will normally have won an award demonstrating their commitment to excellence. The award that exists in the UK is the EFQM (European Foundation for Quality Management) Excellence Model (described in Chapter 7). The working group who wrote the *KPI Report for The Minister for Construction* state that:

> ... the most effective tool for analysing all aspects of an organisation's operations is the EFQM (Business) Excellence Model promoted in the UK by the British Quality Foundation. This enables comparison with other firms and other industries. (DETR, 2000: p. 8)

The remainder of this book is dedicated to explaining how any organisation can develop sufficient expertise and competence to use the principles of benchmarking to implement improvement comparable to organisations that are acknowledged as being the best in the world.

1.7 *A brief outline of subsequent chapters*

- Chapter 2 explains what the concept of benchmarking involves, and how by using it, an organisation can attempt to produce radical improvement.
- Chapter 3 describes TQM and why any organisation that wants to implement benchmarking should understand its underlying philosophy and principles.
- Chapter 4 considers the issue of cultural change in organisations; in particular, the aspect of leadership towards producing such change is addressed.
- Chapter 5 explains what critical success factors and key performance indicators are. In order to use these, it is important that an organisation must understand the day-to-day processes that are used, and how they can be improved by being incorporated into a quality management system.
- Chapter 6 describes what can be regarded as the core concept of benchmarking, namely how to ensure that the satisfaction levels of customers are continuously measured and improved. The concept of relationship marketing and its importance in developing customer loyalty is explained.
- Chapter 7 considers what being acknowledged as 'world class' involves; in particular, how the EFQM Excellence Model can be used to attempt to achieve such status is explained.
- Chapter 8 will explain how, having understood the theoretical aspects of benchmarking, managers can implement the technique in their own organisation. In order to assist them in doing this, a number of case studies that have been written by practitioners using benchmarking in construction are presented. Having read these case studies it is hoped that readers will be inspired to implement benchmarking in order to achieve best practice. As a result employees and customers can enjoy the sort of benefits that normally accrue to organisations that continually strive to be regarded as the best in the world.

CHAPTER TWO

GETTING TO GRIPS WITH THE CONCEPTS

Objectives of this chapter

The previous chapter dealt with the influences that are currently being placed upon the construction industry to promote change. This chapter aims to demystify the terminology that is used in connection with the techniques involved. As such, it will:

- Explain the terms 'benchmarking' and 'best practice'
- Describe the three main types of benchmarking
- Describe the development of benchmarking as a management tool for improvement
- Explain why, because of a serious risk of losing business, Rank Xerox developed the management technique of benchmarking

2.1 *Establishing the principle of benchmarking for best practice*

The terms 'benchmarking' and 'best practice' are, some might suggest, words that are currently fashionable, and will therefore soon be discredited. As this book explains, whilst many of the techniques directly associated with benchmarking and best practice may have their roots in the so-called quality movement of the 1980s, the concepts which underpin them are a lot older. Indeed, their origins are particularly pertinent to construction, as Sylvia Codling in her book *Best Practice Benchmarking* (written for a non-specific audience) explains:

> Records [show that] the Egyptians used benchmarks in construction work [by cutting] a notch in a lump of stone at

> accurately determined points, while a flat strip of iron would then be placed horizontally in the incision to act as the support (bench) for a levelling staff. Using this as the reference (mark) further heights and distances could be measured. (Codling, 1992: p. 1)

As Codling accepts, whilst the term benchmark may have assumed a more contemporary meaning, at its heart is still the fundamental principle of being able to measure in a definitive way. As I will describe, benchmarking - a term now associated with management and business - means being able to measure relative performance by whatever criteria are found to be useful. However, this is nothing new. Businesses have been doing this for as long as commercial enterprises have existed. Best practice is no different. There have always been ways of doing things that were regarded as either superior or more efficient. As the master craftsmen knew, closely guarding the secrets of their trade was an essential way of maintaining their prestige; the only way to learn their 'best practice' was to join them.

The principle of benchmarking and best practice in construction is based upon the assumption that there are usually a number of approaches to carrying out any task, and that these tasks involve certain processes (the significance of the word 'process' will be explained in Chapter 5). Thus, if any individual or group wishes to consider how it should attempt to improve the way it carries out any task(s), the best method is to look at how others do so. From there, it should then be possible to consider how current processes can be changed. However, how can one know if the changes have made any difference? This is why measurement is essential. Like any athlete, unless you have a definitive measure - like the time it takes to run a certain distance - it will be impossible to monitor the effects of alteration.

While carrying out previous research into quality management in construction, and in particular, quality managers (McCabe, 1998), I discovered that one of the main difficulties facing those attempting to introduce quality management into construction organisations is the perception that nothing new can be learned. As the quality managers I studied frequently admitted, one of the biggest challenges they faced was in getting people - at every level of their organisations - to continuously challenge the established ways of doing things and to learn lessons from elsewhere. In organisations where TQM (Total Quality Management) was apparently being effectively implemented, it had been possible to convince some employees that they should reconsider how they carried out their day-to-day tasks. As described later in this book, the effect of doing this enabled people to attempt to implement alterations to the processes involved, the result of which, in many cases, resulted in some measurable improvement. The previous chapter suggests that the report *Rethinking Construction* (Construction Industry Task Force, 1998) argues that benchmarking, in order to learn best practice from elsewhere, is an essential part of improvement activities in construction.

Learning what the principle of benchmarking represents for best practice and how it may be used in any construction organisation, regardless of size, is the main objective of this book. My hope, therefore, is that having read about what the technique involves, readers will feel sufficiently confident in implementing it within their own organisation. With this in mind, I have endeavoured to keep the language simple and to the point. Remember, every successful organisation – particularly those that have achieved what is called 'world class' status (a term that will be explained in Chapter 7) – had to start somewhere. A common feature that is apparent from the reports of these organisations is the desire to ensure that the satisfaction their customers derive from their products or services is continuously being improved. As will be stressed throughout this book, in order to continuously satisfy their customers, such organisations are not afraid to learn best practice from elsewhere. It is worth pointing out that Dr W. Edwards Deming – the person many argue was directly responsible for the Japanese quality revolution (see Chapter 3, section 3.2) – continually stressed that 'the customer is the most important part of the production line' (Walton, 1989: p. 14).

2.2 *Defining benchmarking and best practice*

When one consults texts on the subject of benchmarking and best practice, every author provides their own definition. Most tend to be similar. However, one in particular neatly summarises all the essential features of benchmarking. Thus, according to McGeorge and Palmer, benchmarking is:

> A process of continuous improvement based on the comparison of an organisation's processes or products with those identified as best practice. The best practice comparison is used as a means of establishing achievable goals aimed at obtaining organisational superiority. (McGeorge & Palmer, 1997, p. 83)

What stands out in this definition which, incidentally, comes from a book written for construction, is that it uses the words 'continuous improvement' right at the start; there can be no doubt as to their importance. Indeed, as McGeorge and Palmer explain, the main purpose of benchmarking is the 'search for best practice', not, they suggest, merely to indicate that what you currently do is adequate. However, the difficulty that is often encountered in the early stages is to know what best practice is, and more importantly, where it exists. As the following section describes, there are various ways of carrying out benchmarking which seek to discover best practice within your own organisation, in competitors, and in organisations whose products or services may be completely unrelated to yours. Fig. 2.1 shows what McGeorge and Palmer suggest to be an inverted 'pyramid of success' (McGeorge & Palmer, 1997: p. 82)

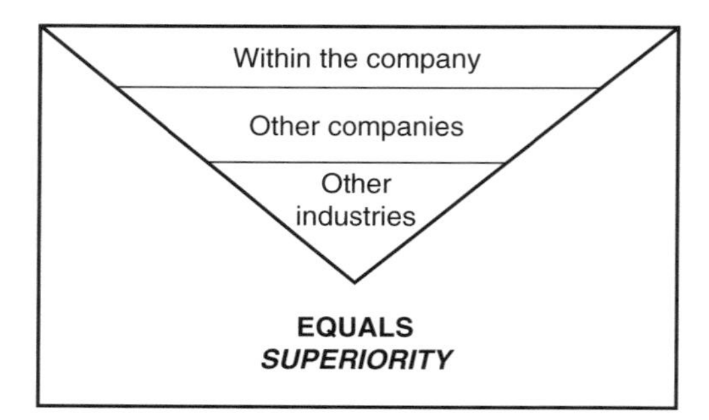

Fig. 2.1 The objectives of success using benchmarking.

Another word which McGeorge and Palmer use which is crucial to successful benchmarking is *process*. As Bendell *et al.* explain, 'processes are mechanisms by which *inputs* are transformed into *outputs*' (Bendell *et al.*, 1997: p. 92). Thus, in every business – regardless of size or type of business – there will be processes going on day in, day out. They will be carried out in order to allow the organisation to achieve its aims. However, and as will be explained later, one of the biggest challenges that any organisation faces is being able to identify what all of the processes are, and how they interrelate. It is frequently the case that whilst individuals know what their own processes are, they do not know those that relate to other activities. The need to achieve quality assurance (ISO 9000; BSI, 1994) has meant that processes are now more likely to be formally documented than would have been the case in the past.

However, if improvement is to occur it may be necessary to consider altering some of the processes, or change the way that different processes interact with one another. As Bendell *et al.* describe, benchmarking is likely to assist in providing guidance as to what changes should be initiated in an organisation's processes (Bendell *et al.*, 1997).

2.3 *Types of benchmarking*

There are three main ways of carrying out benchmarking:

- Internal
- Competitive
- Functional or generic

2.3.1 Internal benchmarking

Of the three methods to be discussed, internal benchmarking is the most straightforward. This is because, as the title implies, the activity is carried out inside the organisation. In looking at its own processes, it may be discovered that a particular department is able to perform more efficiently than others. There may be many reasons for this. It may be that the way that activities are being carried out in this department has been achieved because the processes have been engineered to ensure efficiency is the primary objective. Alternatively, there may be a good team spirit which allows people to co-operate. The challenge to other departments therefore is to consider whether this 'best practice' could be transferred.

Because internal benchmarking does not require access to other organisations, it is unlikely to mean that there will be any reason why best practice cannot be used to assist other departments. Indeed, it would be a strange organisation which did not do everything possible to encourage this. Moreover, the fact that departments are encouraged to communicate with one another will be highly likely to assist in enabling people to understand what goes on elsewhere. There is a tendency for people to think that only their own task is important; how others do their tasks is their problem. The fact that in order to do their tasks they rely on the efforts of others does not seem important. The reality is more usually that overall improvement can only occur if every person/department in an organisation collaborates and works as a team. Think of the word team as an acronym for '*Together each achieves more*'. Moreover, as Bendell *et al.* explain, 'the boundaries of each part of the processes [are a] key element in the implementation of TQM' (Bendell *et al.*, 1997: p. 93).

2.3.2 Competitive benchmarking

Any business organisation which is in direct competition with others will be presumed to monitor what those others do, and more importantly, how. Clearly, if a competitor appears to have suddenly gained a competitive advantage - such as being able to sell its goods cheaper or to a higher specification - other companies will probably be forced to follow suit. As the Rank Xerox story demonstrates (see section 2.4), the likely result of not matching competitors is an inability to sell; the logical consequence of which is to be forced out of business. Competitive benchmarking therefore is based on attempting to compare your processes with organisations that produce and sell the same goods or services as you, particularly those with commercial advantages.

Competitive benchmarking has certain problems. The first is obvious: competitors are hardly going to tell you how you can beat them. It may be necessary to be prepared to travel to organisations which, whilst being in the same business, do not operate in the same geographical region. Going abroad may be a good way of seeing how others do things (something the CIOB did in Japan (see Chartered Institute of Building, 1995)). The other problem of competitive benchmarking is that comparison against others in the same sector may not result in the belief that radical changes are required in your processes. In construction, it is rare for organisations to consider doing things very differently to their competitors. However, as the case studies in Chapter 8 show, there are some construction organisations which have begun to implement radical changes to their processes. Notably though, some of these organisations did so after having used the last of the methods of benchmarking, i.e. functional or generic.

2.3.3 Functional or generic benchmarking

Functional or generic could be regarded as the form of benchmarking that is likely to result in the most change in an organisation's processes. The reason for this is that you are attempting to compare your processes against those of organisations which are considered to be the 'best in class'. So, for instance, you might consider going to one of the Japanese car factories that operate in the UK. What you are likely to see in an organisation which is regarded as being 'world-class' (a concept discussed in greater detail later) are processes being carried out which result in efficiencies greater than those of their competitors, and extremely high levels of customer satisfaction.

This type of benchmarking has the advantage that you are not trying to learn from those against whom you are competing. It is also notable that organisations which have achieved the accolade of being 'world class' or award-winners such as the EFQM (European Foundation for Quality Management) (see Chapter 7 for detailed explanation), are prepared to share the secrets of their success. The reason for this is believed to be their confidence that they are so far ahead that even their direct competitors are not capable of catching them. The only problem that this form of benchmarking may have, is knowing how to implement what you learn from these organisations in your own. As someone once told me, 'There is a danger of engaging in industrial tourism'. This means that you should have carried out internal benchmarking of some key processes prior to carrying out the functional or generic type. It is worth remembering that every organisation, regardless of the context in which it operates, will have certain processes for dealing with things like buying materials, managing its workforce and, most importantly, dealing with its customers.

The main advantage of this form of benchmarking is that the objective is to be inspired to attempt to do things differently if it will ensure improvement. Indeed, as McGeorge and Palmer assert:

> ... it breaks down barriers to thinking and offers a great opportunity for innovation. It also broadens the knowledge base and offers creative and stimulating ideas. (McGeorge & Palmer, 1997: p. 88)

McGeorge and Palmer provide a very useful diagram that summarises the main aspects involved in the use of either generic, competitive or internal benchmarking (Fig. 2.2). As they explain, when moving from internal to generic, 'the level of difficulty, the time taken and cost incurred increases along with the creativity and the opportunity for improvement' (McGeorge & Palmer, 1997). However, as they also explain, when moving from generic to internal, 'cost, time and difficulty decrease as do relevance, ease of

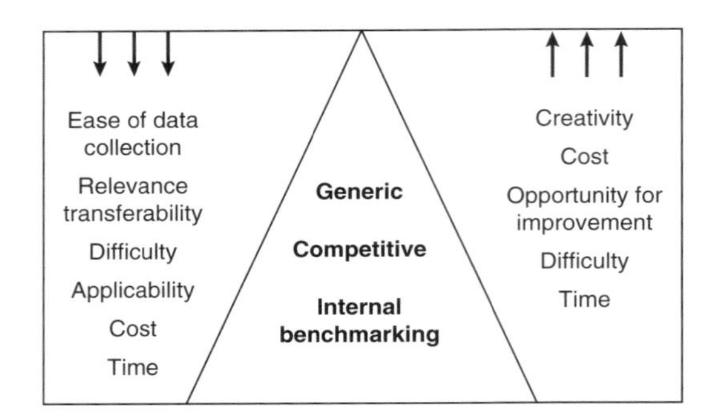

Fig. 2.2 The relative advantages of different types of benchmarking.

data collection, applicability and transferability of results' (McGeorge & Palmer, 1997).

Learning and knowledge are key themes which this book will return to many times. They are essential in creating the right environment in an organisation (commonly referred to as its culture). After all, what is the point of seeing how the best in the world carry out their processes if employees are unable or unprepared to change? The need to create an environment where learning and willingness to change are issues which will be returned to many times within this book.

At this point, it is worth considering the Rank Xerox story. It is seminal in so far as its shows that an unanticipated commercial threat can, if dealt with by sensible benchmarking, result in radical change that will not only save the business, but eventually lead to achievement of superiority in the market you operate in. As a consequence, what Rank Xerox achieved using benchmarking has, in effect, become the benchmark against which organisations that wish to radically improve measure themselves.

2.4 The Rank Xerox story

The word 'benchmarking' can be credited to Robert Camp (1989) who wrote the book *Benchmarking: The Search for Industry Best Practices that Lead to Superior Performance*. In this book, Camp describes why, in 1979, Rank Xerox in America realised, after a decade of rising costs and reducing market share in photocopiers, that they needed to radically change the way that operations were carried out. The most persuasive thing that forced Rank Xerox to consider change was that they became aware that Japanese competitors were capable of competing by producing copiers both better and cheaper. Because of a combination of both dominance and complacency, Rank Xerox had never considered the Japanese to pose a threat. Indeed, as Cross and Leonard explain, Rank Xerox, following a study of the Japanese competitors in the photocopier market, discovered that, as well as being able to sell them cheaper, the product quality was actually better:

> The results of benchmarking were startling. Costs were too high ... [Xerox's] unit manufacturing cost was the same as their competitors' selling price. The assumption prior to benchmarking was that the competitors' machines were poor quality. This was proved by benchmarking to be wrong and to drive the point home they were making profit! (Cross & Leonard, 1994: p. 498)

Clearly, Rank Xerox realised that there was the very definite potential for commercial destruction; if your customers can buy elsewhere what you sell, and cheaper and better, why should they continue to purchase? As Cross and Leonard assert, the introduction of benchmarking and continuous improvement has resulted in Rank Xerox moving from being an organisation which was 'internally focused', to one which has 'one objective: 100 per cent customer satisfaction' (Cross & Leonard, 1994).

2.4.1 What did Rank Xerox do?

The most obvious benchmark that Rank Xerox were aware of was price; any organisation which provides products or service for profit or non-profit will be aware of what its competitors charge. In a world where everything is assumed to be equal, price is a good way to make judgements (this, of course, is the basis upon which competitive tendering is carried out). However, as we are aware, price is not the only measure that customers use for decision-making. This, Rank Xerox quickly discovered, was the reason for losing customers; besides being able to buy cheaper Japanese products, buyers could get a copier that would perform better. From the intelligence that could be gathered from what amounted to competitive benchmarking of its Japanese competitors, it was discovered that Rank Xerox:

- had nine times as many suppliers
- produced 30 000 defective parts per million (the Japanese equivalent was 1000)

- took twice as long to get its products from design to market
- used five times as many engineers
- implemented four times as many design changes
- had design costs that were three times higher
- rejected ten times as many machines on the production line

These facts, despite causing alarm, did at least provide Rank Xerox with targets against which to measure its improvement (as I have already explained, this is the most fundamental component of benchmarking). What Rank Xerox had to do as a result, was to find ways by which to improve its processes. Fortuitously, Rank Xerox had the advantage of being able to compare itself 'internally' with Fuji Xerox – a Japanese joint venture – in order to find out why it was performing so badly against all of these benchmarks. What quickly became apparent was that Fuji Xerox, like many other manufacturers in Japan, had wholeheartedly taken on board the principles that quality gurus such as Dr Deming had recommended after World War II. Additionally, Rank Xerox used functional/ generic benchmarking against an American mail-order company, L.L. Bean – regarded as a 'best practice' company with regard to productivity – in order to benchmark its stock-keeping function. Karlof and Ostblom explain that despite the fact that Bean's business was unrelated to that of Rank Xerox, 'the logic of the order processing, stock-keeping and invoicing routines was the same regardless of product' (Karlof & Ostblom, 1993: p. 48). By looking at what L.L. Bean did with respect to stock-keeping, Rank Xerox was able to learn lessons which allowed it to immediately change its own processes.

Benchmarking by organisations is based upon the assumption of being prepared to implement total quality, a concept that will be fully described in the next chapter. Prior to that, it is worth stating that total quality management assumes that many things will occur, most especially employee involvement. However, unless there is a visible commitment by senior management to any initiative, its chances of success are doomed. In this regard, Rank Xerox had as its chief executive (after his appointment in 1982) David Kearns, an enthusiastic proponent of the principles of TQM. As Kearns and Nadler (1992) describe in *Prophets in the Dark: How Xerox reinvented itself and beat back the Japanese*, it was decided that there should be a five-year plan involving 18% growth year on year to catch up with the Japanese, and that this was called 'Leadership Through Quality'. In carrying out research into the use of quality management, the need for senior management commitment is essential. After all, unless they are prepared to support everyone's efforts, and in particular, provide adequate resources, it will be impossible for employees below them to ensure success, no matter how well-intentioned they are.

Rank Xerox have summarised their approach to benchmarking by the use of a five-phase/ten-step model. This model, which is described in detail by Camp (1989), is shown below.

Phase One: Planning

(1) Identify what aspects of the organisation's outputs and/or processes should be benchmarked
(2) Identify the organisation which is regarded as your 'best competitor'
(3) Consider how best to collect data which will be capable of being used meaningfully

As well as the need for senior managers to show their commitment to the effort, Rank Xerox also stress that it is a teamwork exercise. As a result, they explain, it is essential that at this phase there is total agreement on how the exercise will be carried out. Moreover, it is recommended that the team consists of employees from all hierarchical levels of the organisation. Crucially, Rank Xerox believe, benchmarking provides an extremely valuable opportunity for senior managers to understand what those at operational level do, and more importantly, what they need in order to do a better job. By implication, the issue of organisational culture will need to be addressed - something I will describe in detail in Chapter 4.

Rank Xerox recommend that in order for the benchmarking initiative to succeed, there must be a careful selection of attempting to do things which are likely to be achieved. In particular, they warn against the desire to try and change everything at once. The danger of failure, whilst always being present, must be avoided; that will cause demotivation of those involved. It is worth stating that Rank Xerox do not believe that improvement occurs simply by hoping that people will be willingly involved; there is a need to both convince and train those who contribute. At the start of every Rank Xerox training session, the advice that Sun Tzu offered to Chinese warlords 2000 years ago is quoted. This is, 'If you know your enemy and know yourself, you need not fear the results of a hundred battles'. The message that Rank Xerox wishes to give to its employees is that in order to beat the opposition, it is essential to know your own weaknesses.

Phase Two: Analysis

(4) Find out how far behind you are; known as the 'competitive gap'
(5) Decide on the levels of performance that you wish to reach in the future

The ability to carry out analysis, Rank Xerox believe, is dependent upon the quality of the information collected as a result of phase one. The main emphasis is upon eliciting objective data. Clearly, the better the effort is thought out, i.e. planned, the greater the chances are of being able to consider how to improve your internal processes. It is for this reason that attempting to deal with one generic process at a time is very prudent; as, for instance, Rank Xerox did with respect to stock-keeping in L.L. Bean. What is most important at this stage is to be able to determine the 'competitive gap' between what you do and the organisation against which you are benchmarking. Like Rank Xerox, an organisation which benchmarks wishes to know how far behind it is, and how quickly it can catch up, i.e. 'close the gap'. Thus, decisions can then be considered as to how to deal with step five, the future performance levels you want to attain in the future. It is worth pointing out that whilst considering how to catch up with competitors, particularly those who are regarded as the 'best', there is a strong likelihood that their rate of improvement will be faster than you are currently capable of. The realisation of the magnitude of how much effort is required may cause concern at this point, perhaps even, a temptation to give up. The whole point of benchmarking, it must be remembered, is to have objective evidence of how uncompetitive you are. As Rank Xerox discovered, unless you know the degree of risk you are at from others, it is impossible to deal with it.

Phase Three: Integration

(6) Establishment of the functional goals
(7) Put into place plans which will achieve these functional goals

This is the phase where it is necessary to consider what will be done, how, and by whom. At this point, every person who will be involved - assuming they were not part of the teams which carried out phases one and two - will need to be adequately informed of what is happening and why. As will be described in greater detail in Chapter 4, a major part of producing cultural change is the need to encourage a willingness to engage on the part of participants. If any change is forced upon people, it will be unlikely to be welcomed, and as a result, resisted. A key element of so-called 'excellent organisations' – such as Rank Xerox – is the fact that the people they employ are actively involved in the major decision-making. As a consequence, change management leading to improvement is far less traumatic.

Phase Four: Action

(8) The introduction of changes in processes
(9) Monitor results of the changes to identify areas where improvement has, or has not, occurred
(10) Consider what has been achieved and, if necessary, 'recalibrate' the benchmarks and ensure that feedback occurs prior to the next phase of planning

It is at this point that things should start to happen. All of the previous phases will have identified the processes that need to be improved and by what means. In exactly the same way as benchmarking against others is about comparing (measuring) relative performance, so introducing change is carried out on the basis of measuring its effects. What can be measured and how is the subject of Chapter 5. However, whilst the aim is to improve, if changing the way a process is carried out leads to a reduction in whatever measure is used, the most sensible thing to do is stop and investigate why. If necessary, consider changing the benchmarks or ways that the processes are being altered. A theme that this book will continue to return to is the need for those involved in benchmarking to be aware of the effects of what they are doing and be prepared to continuously learn. It is for this reason that Rank Xerox continuously 'recalibrate' the benchmarks.

Phase Five: Maturity

There are no steps to this phase. The objective is to have created an organisational environment where, according to Cross and Leonard (1994), the process of benchmarking has become fully integrated into all the practices of the organisation. In effect, and as Rank Xerox have shown by their experience, every person or team is continously striving to create ways to improve what it does on a day-to-day basis. The result of this combination of effort, and it should be stressed *culture change*, is to allow organisations such as Rank Xerox not only to regain market position, but also, by continuously focusing on customer satisfaction, to be regarded as one of the 'best in their class'.

In effect, Rank Xerox's experience has shown that it is possible to use benchmarking not only as a tool to survive, but also as one by which to set standards which others will wish to emulate. However, as the next chapter describes, the philosophy of benchmarking and TQM requires that the organisation – or most especially its *people* – must be willing to continuously learn how to do everything better; a concept known as *Kaizen*.

2.4.2 Summarising the Rank Xerox approach

Rank Xerox has, because of the experience of losing customer share, dedicated itself to ensuring that all of its activities are carried out with the sole intention of creating continuous improvement in the satisfaction levels of its customers (see Fig. 2.3).

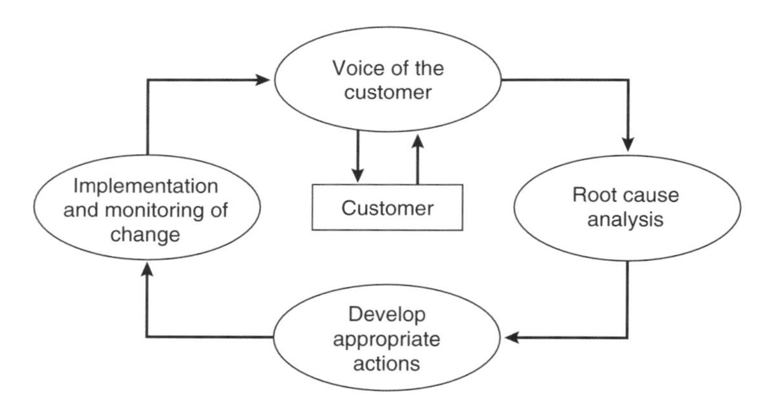

Fig. 2.3 The Rank Xerox customer orientation model.

In order to produce a 'total quality company', Rank Xerox believes that it is essential to bring together six essential elements. These are:

(1) Reward and recognition for all employees who demonstrate their willingness to create opportunities for change
(2) Processes and tools that achieve the overall objective of improvement (see Fig. 2.4)
(3) Training to all employees in the use of tools and techniques

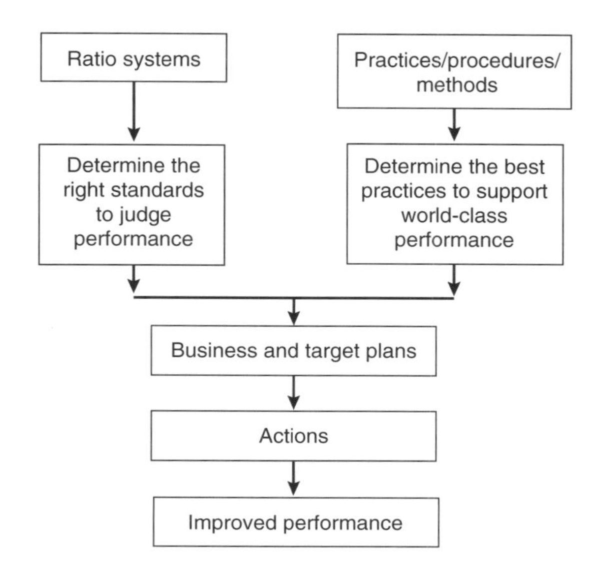

Fig. 2.4 Tools and techniques in use in Rank Xerox.

(4) Transition management to achieve change
(5) Communication of philosophy to ensure clarity in objectives to all
(6) Management behaviour that is seen to be absolutely committed to improvement

2.5 Conclusion

In what is probably the most influential British book to be written on the subject of TQM, Oakland suggests that there are five key reasons why an organisation should engage in benchmarking (Oakland, 1993: p. 181). These provide a useful synopsis of the principles that have been described in this chapter:

(1) *Becoming competitive.* As Oakland argues, if an organisation does not look to what is going on elsewhere, it will miss out on understanding what has been proven to work. Moreover, he believes, without benchmarking, any initiatives attempted will tend to be both 'internally focused [and] evolutionary' (Oakland, 1993). The overall consequence of not wanting to be competitive hardly need emphasising!

(2) *Industry best practice.* Oakland asserts that an organisation can only benefit from learning what is regarded as best practice. As

he explains, as well as providing a wider range of options for doing things, it will allow the organisation to achieve a more superior performance.

(3) *Defining the customer.* This book has already established the centrality of the customer in defining an organisation's objectives. Oakland, like all other advocates of benchmarking, believes that the need to provide more 'objective evaluation' of customer requirements is fundamental to the continued ability to successfully satisfy their desires.

(4) *Establishing effective goals and objectives.* In attempting to achieve any objective, it is essential to have a clear vision of the desired outcome. In benchmarking, according to Oakland, unless there are effective goals and objectives that are 'credible, unarguable [and] proactive', it is less likely for the initiative to be successful. The alternative to having clear goals will be for any improvement initiative to be lacking in focus and simply 'reactive'.

(5) *Developing true measures (of cause and effect).* One of the principles that this book will constantly stress, is the need for measures to be put in place which allow assessment of data which are produced in respect to any changes made. Without such measures, the effects that are brought about will tend to be purely anecdotal. As a result, it will be impossible to make an objective judgement as to how effective the changes were.

Summary

This chapter has described the following:

- What benchmarking and best practice mean
- How these concepts can be used as the basis upon which improvement can occur
- The three types of benchmarking that are used
  - Internal
  - Competitive
  - Functional/generic
- Where the concept of benchmarking originated from – the Rank Xerox story – and the five-phase/ten-step model they recommend

CHAPTER THREE
WHAT IS TQM AND ITS IMPORTANCE TO BENCHMARKING?

Objectives

Chapter 2 explained what benchmarking is, and the potential it provides as a management tool for organisational improvement. In providing this explanation it was suggested that certain things needed to accompany the benchmarking effort, most especially the development of a culture in which people are encouraged and supported to engage in activities that allow continuous improvement in all aspects of the business. In effect, such an approach is usually generically called TQM (Total Quality Management). Therefore, this chapter will provide the following:

- Describe the fundamental principles of TQM
- A short history of the development of quality in Japan in the post World War II period - the influence of Deming and Juran
- How the philosophy of TQM is used to produce continuous improvement
- The central principle of the customer in TQM

3.1 *Defining TQM*

The task of defining TQM is similar to that of benchmarking; whilst all tend to include the same features, many authors choose to interpret each of them slightly differently. This, I believe, is not a particular problem as long as certain principles are included, the most important of which is the involvement by all people in the process who contribute to ensuring customer satisfaction. It should therefore be no surprise that a definition provided by the British Standards Institution, in BS EN ISO 8402, does exactly this:

> TQM is a management approach, centred on quality, based on the participation of all members and aiming at long-term success through customer satisfaction (BSI, 1995: p. 27)

As the BSI definition explains, as well as the objective of ensuring customer satisfaction, there will be benefits to all of the 'members of [the] organisation and to society'. In the next section, I describe the influence of two quality 'gurus' – Dr Deming and Dr Juran – both of whom are credited with having assisted the Japanese to rebuild their industrial capacity after World War II. Despite the recent economic difficulties that Japan has suffered, there can be little argument as to the dominance which Japanese producers of electronic and automotive goods have enjoyed for the last 20 years. As will be explained, Deming told those senior managers who attended his lectures in Japan that improvement of quality would result in reducing costs, increased efficiency and less wastage. This, he argued, would enable them to provide customers with products which, as well as being cheaper, would be less likely to fail (thus increasing customer satisfaction). As a consequence of doing this, he asserted, Japanese producers would increase their markets, improve their profit levels, and have the funds to invest in further product development which would enable them to attract even more customers. This explanation, referred to as the 'Deming Chain Reaction' has the effect of not just securing jobs for those already employed, but of providing the potential for additional employment. As the Japanese discovered by following what are now referred to as TQM principles, they were able to provide secure (lifelong) employment to their employees and help society to enjoy rapidly increasing standards of living. Indeed, as will be shown in Chapter 7 the EFQM Excellence Model – a derivative of The Deming Prize, and which is regarded as a way of recognising (benchmarking) those organisations which have produced superior levels of quality – places emphasis upon the need for organisations to consider the impact of their actions on society.

3.2 The origins of TQM – the influence of Deming (1900–1993) and Juran (1904)

In describing how post World War II Japan achieved pre-eminence in terms of quality, this book has alluded to the influence of Dr W. Edwards Deming, an American statistician. Whilst it could be argued that Deming was the most influential of the management advisers that visited a defeated Japan under the control of General MacArthur, Supreme Commander of the Allied Forces, Dr Joseph Juran deserves as much credit. For reasons that are described more fully elsewhere (McCabe, 1998), there has been a tendency to forget the work that Juran did in Japan.

The history of how Japanese industry recovered from World War II is both instructive, and seminal to the subject of benchmarking. As Morrison is moved to state, '[Japan has been able to] achieve, by peaceful economic means, what they had failed to do by war: to dominate the World' (Morrison, 1994: p. 52). In the aftermath of war, Japan's industrial capabilities were ruined. Moreover, as a form of punishment for their aggression, MacArthur demanded that all senior and middle managers in their factories should be sacked and younger junior managers put in charge. Given what was to happen later, this act alone might be regarded as having provided a fertile 'ground' upon which Deming and Juran could 'sow the seeds' of the new philosophy that they preached. Those who now controlled Japanese industry, in which previously the principle of feudalism (age) had been used as the only method of promotion, lacked experience in senior management. If the only advice they received came from American experts, then so be it. As Walton remarks, the managers who found themselves in control of post World War II Japanese industry, 'having lost all, had nothing to lose' (Walton, 1989: p. 14).

3.2.1 SPC (Statistical Process Control) – the cornerstone of Deming's philosophy

In describing Deming, and in particular his use of what is known as 'SPC' (Statistical Process Control), it is necessary to refer to the person who originally developed the concept and acted as his early mentor, Dr William Shewhart. In what is a fascinating coincidence Shewhart, Deming and Juran all worked at the same factory – the Hawthorne Factory – the place which, according to Kennedy, is now regarded as the 'crucible of influential research [into early theories of motivation and as a result] going down in industrial history as the source of industrial sociology' (Kennedy, 1994: p. 217). This may appear to be just a piece of interesting, but not significant trivia. However, whilst it is does not seem that either Shewhart, Deming and Juran were directly influenced in the experiments carried out by Elton Mayo[16] *et al.*, the message that both Deming and Juran emphasised to their Japanese audiences was the importance of people in producing quality. This raises the issue of culture, something that will recur many times in this book.

The emphasis upon how to combine the systems by which people operate, and how they co-operate, is at the heart of SPC. SPC, because of its reliance on statistics, can appear daunting. However, using simple statistics in order to carry out measurement - something that is crucial to the philosophy of benchmarking - is, as Deming continuously stressed to his Japanese audiences, essential. The fact is, any process, regardless of how relatively unimportant, can be measured in terms of time or output. Therefore, Deming advised, subsequent to having conducted the measurements, and using simple formulae, it is possible to plot a control chart (including control limits) that shows the variation of the measurement over time. A typical control chart is shown below in Fig. 3.1

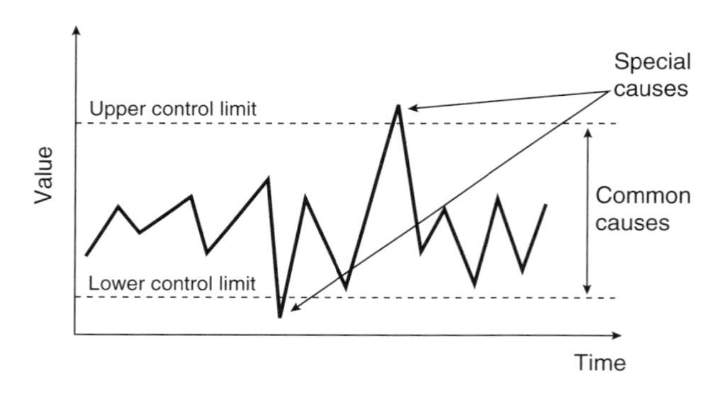

Fig. 3.1 Typical control chart.

The aim of using SPC is to continuously improve the process so as to reduce variation. The less variation there is, Deming explained, the more the process is 'under control' and therefore likely to succeed in producing the intended results. What is so important about SPC, however, is to recognise why the variation is occurring, and as a result, to correctly assign responsibility. As Fig. 3.1 shows, the variation can either be within the control limits, i.e. it has *common causes*, or it goes beyond the control limits, i.e. it has *special causes*.

This distinction between common and special causes is of absolute importance because, as Shewhart and Deming advised, special causes are usually highly irregular (less than 10%), and due to things that only the person(s) most directly involved in the process can know about. Common causes are those which provide most of the variation (90% plus), and are due to the way that the process has been set up. Crucially, as Shewhart and Deming stressed, only the managers who are responsible for controlling the way in which systems have been designed to control processes, can create the conditions by which most variation occurs and conversely can cause improvement to happen. As Deming explained, this avoided the temptation by managers to blame all problems on workers. Instead, he advised, it is the responsibility of managers to constantly strive to improve the overall system. Unless they do so, he believed, simply expecting workers to create improvement will not give the customer what they want. Moreover, as well as changing their attitude to workers, Deming recommended that managers should be prepared to work more closely with them in managing the processes to produce customer satisfaction. This shift in behaviour and treatment of workers - what is now commonly referred to as cultural change - was willingly accepted by the Japanese managers to whom Deming gave his lectures. As Crainer comments, with respect to the changes which occurred in Japanese work practices as a result of Deming's advice:

> Deming appreciated that no matter how powerful the tool of mathematical statistics might be, it would be ineffective unless used in the correct cultural context. (Crainer, 1996: p. 143)

Crainer believes that it was the combination of 'culture and measurement', which the Japanese used to such powerful effect in dominating the electronic and automotive sectors, that led the Americans to 'rediscover' Deming and invent 'what is now labelled Total Quality Management' (Crainer, 1996).

Deming's work is believed to be enshrined in what are known as the 'Deming Chain Reaction' and 'The PDCA Cycle'. The former suggests that once improvement takes hold in an organisation, it becomes a virtuous cycle. The latter summarises the desire to continuously improve each and every process. These rationales are shown in Figs 3.2 and 3.3 below.

3.2.2 Juran's quality trilogy

Juran is very similar to Deming in many ways: he also worked with, and was influenced by, Shewhart at the Hawthorne plant; he lectured to the Japanese about quality in the period immediately after World War II. However, the approach to improving quality

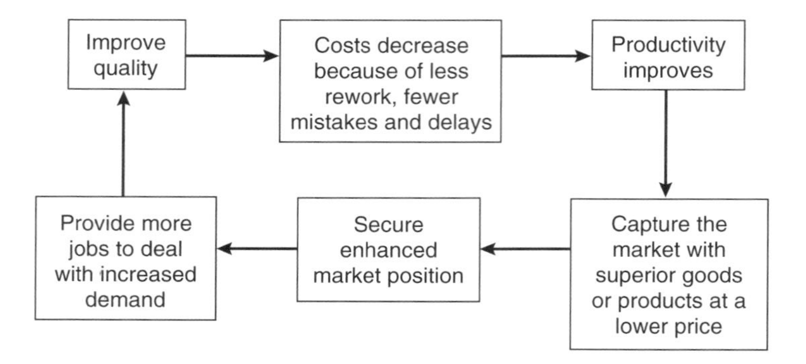

Fig. 3.2 The Deming Chain Reaction.

that Juran proposed was different to Deming's in the following ways:

(1) *Deming concentrated on education; Juran believed that the key to success lay in actual implementation.* As a result of this, Juran advanced what he called Company-Wide Quality Management (CWQM), the essence of which is that improvement must come from every level of the company. It is for this reason that whereas Deming insisted on lecturing to the *Kei-dan-ren* (the association of Japanese chief executives), Juran tended to concentrate his message on those who he considered to be in the best position to actually influence improvement, namely middle management and quality professionals. This is an extremely important point. As will be explained in Chapter 4,

Fig. 3.3 The Deming Plan, Do, Check, Action (PDCA) Cycle.

the author's research has shown that those who 'champion' the cause of total quality/improvement/benchmarking are usually from the ranks of the middle managers.

(2) *Juran stressed that the key to managing improvement was to concentrate on the cost of quality.* Deming was vehement in his belief that management in the West tended to be obsessed with cost (something he summarised as 'running a company on visible figures alone - counting the money' in his 'seven deadly diseases'). Juran did not believe that this was a problem. As Dale, *et al.* state with respect to Juran's emphasis on cost, 'the language of top management is money' (Dale *et al.*, 1994: p. 19). Thus, as Juran explained, if senior managers can see an improvement in the bottom line because of the efforts being made to deal with quality, they will be likely to support further efforts.

Juran's philosophy is summarised as involving four steps which lead to a 'quality trilogy'. These are:

(1) *Step one* - clearly identify specific things/projects that need to be done
(2) *Step two* - provide definite plans for achieving what can be done
(3) *Step three* - ensure that people are made responsible for doing certain things
(4) *Step four* - make sure that the lessons that are learned during the previous three steps are captured and used in feedback

The quality trilogy (see Fig. 3.4) involves three essential aspects of quality:

(1) Planning
(2) Control
(3) Improvement

These are now discussed in turn.

Planning

Juran was adamant that the ability of an organisation to produce quality can never happen by accident; it must be planned for (this surely is a principle that everyone who works in construction will

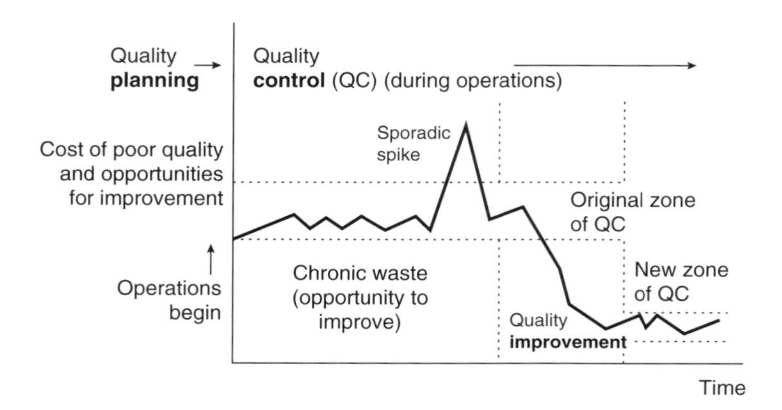

Fig. 3.4 Juran's quality trilogy.

agree with). So, according to Juran, in order to plan for quality, the following four steps are required:

(1) To clearly identify the needs of all of those involved in the process of producing the end-product or service. This is a central tenet of TQM – the need to involve every person/team/department, each of which usually depends on the input of others to carry out their tasks.
(2) To ensure that the needs that are identified in (1) above are put into language which is both simple and unambiguous. This principle is one that we all claim to want, but frequently fail to achieve. As Juran stressed, if we fail to appreciate what others want, we should not be surprised if we are unable to match their expectations.
(3) Having identified needs and articulated them to everyone involved, to develop a process which is both capable and robust enough to achieve the desired outcome, i.e. to satisfy everyone's needs.
(4) To put into place all processes and continually monitor them to ensure continued achievement and improvement which satisfies the customer's needs.

Control

As Juran believed, having put in place processes that are designed to produce improvement, they must be continually monitored – *measured* – to ensure that this indeed occurs. Like Deming, Juran's message is that it is only by controlling the processes that waste can be reduced, and savings made. Also, similar to Deming, Juran, despite his focus on those who are directly in control of operations/ processes, stressed the need for management to be aware of their responsibility in designing a system in which every person can achieve the best result.

Improvement

This is the desired consequence of the other two parts of the trilogy. Once improvement starts to occur, Juran argued, it is incumbent upon senior management to consider ways in which all parts of the organisation can be improved by better systems; the workers, provided they are trained and supported will benefit. This philosophy of improving every aspect of what the organisation does is the reason why the word 'total' is included in TQM.

Overview

Table 3.1 provides a summary of what adopting Juran's philosophy should lead to.

Table 3.1 The likely consequences of using the Juran trilogy

Trilogy process		Likely consequences
Step	**What this entails**	
Planning	Preparation for quality goals to be achieved during production	Process capable of meeting goals
Control	Meeting goals during operations	Operations are occurring in accordance with plans
Improvement	Being able to produce at unprecedented levels of performance	Level of quality achieved surpasses what was believed to be possible

3.3 The Toyota story – an early example of benchmarking

Rank Xerox tends to be regarded as the genesis of benchmarking in order to achieve best practice. However, the way that Toyota came to be regarded as a world-class car producer is a good example of how benchmarking was used in order to learn what bad practices existed, and thus, how to avoid replicating them.

In 1950, a young Japanese engineer, Eiji Toyoda, visited the Ford River Rouge plant in Detroit. The Toyoda family had founded their company in 1937, but, since their name meant 'abundant rice field' they used a contest to decide a more fitting identification. The result, after 27 000 suggestions, was Toyota (a word which has no meaning). The early development of Toyota gave no indication of the fact that in the 1980s and 1990s it would come to be regarded as an exemplar of a company capable of manufacturing cars which, as well as being of an extremely high standard, were produced with levels of efficiency that many Western competitors – until they learned how to use similar techniques – found staggering. In fact, it is amazing to discover that whereas in the 13 years prior to 1950, Toyota had produced only 2685 cars, Ford were producing over three times that number in a single day. It was hardly surprising that the young Eiji Toyoda (a son of one of the founding family) would want to go to the largest and most productive car factory in the world to learn how to improve his own factory.

Toyoda must have been impressed by what he saw at the River Rouge plant. As he would have witnessed at first hand, it relied on mass production systems which, as Pursell believes, 'not only created a great, integrated machine for turning out cars, but pioneered a workforce as standardised and interchangeable as his automobiles and as dedicated to a single purpose as the thousands of machine tools they tended' (Pursell, 1994: p. 99). However, what Toyoda would have also seen for himself was a reduction in the humanity of the worker, as Charlie Chaplin captured in his film masterpiece *Modern Times.* Critically, as Toyoda and a production genius he employed called Taiichi Ohno believed, the system that Ford was using was, despite being capable of producing high numbers of cars, extremely wasteful. It was, they thought, rife with *muda* (the Japanese word for waste) in every aspect of production: effort, materials, time, and most wasteful of all, in sapping the spirit of the workers. As a consequence, Toyda and Ohno were convinced that alternative systems existed which were capable of producing comparable numbers to Ford but with less waste and with a workforce that would be less demotivated than the workers at the River Rouge plant.

The system that Toyoda and Ohno developed – what has become known as lean production – involves certain fundamental principles which are frequently assumed to be synonymous with the philosophy of TQM:

- Flexible working practices (*Shojinka*) by those directly involved in the production process; this enabled workers to control what they were doing
- Teamworking
- Creative thinking (*Soikufu*)
- Because of the ability of workers to change production equipment themselves, cars could be made in small batches which allowed regular changes in the models to rapidly meet changes in customer requirements
- An emphasis on improving the quality of what was being produced (*Kaizen*) rather than Ford's obsession with targets to be achieved

There are other facets of Toyota's production system such as, for instance, closer relationships with suppliers (partners), rationalised stock systems (just-in-time), and rapid development cycles. What is most important to note from the Toyota lesson is that, comparing its processes with Ford, it learned the following two principles and used them to such great competitive effect that, eventually, Western car manufacturers - including Ford - were forced to benchmark themselves against Japanese car producers such as Toyota:

(1) Obsession with customer satisfaction (including direct consultation to develop new models)
(2) Incorporation of participation by workers in a way that could never have been allowed in the Ford (Tayloristic) mass-production system

3.4 The development of TQM in the West

As many commentators argue, had the Japanese simply followed the example of Western producers and sold goods which 'merely satisfied' their customers, interest in emulating the use of methods they had used to such great effect might not have developed. The interest in TQM which initially occurred in the USA is ironical because, as those who examined Japan's apparent success quickly discovered, they had 'simply' rigorously applied what Deming and Juran had taught them. There are many so-called quality gurus who emerged as a result of the need for organisations to radically improve, or risk getting beaten. Describing what these gurus recommend, whilst being interesting, tends not to add a great deal to the message that Deming and Juran taught to the Japanese in the post-war period. What seems obvious, is that the Japanese learned that in order to improve, they must understand their own processes, look at how others carry out similar processes and, using appropriate measures, continually strive to implement more efficient methods of producing goods that achieve customer satisfaction. However, as Crainer (1996) contends, the use of 'hard' statistics must be applied in an organisational environment that is conducive to improvement. This leads to what is often described as the 'soft' side of TQM. It is this soft side, the issues of how people are managed and the environment (culture) in which they carry out their day-to-day tasks, that has become the focus of many of what can be called the 'contemporaries' of quality management (see McCabe, 1998). This focus occurred after the discovery that the application of systems similar to the one that Ford used at River Rouge was entirely misguided, most particularly in the way that measurement was being used to threaten and punish workers.

3.4.1 The move from inspection and quality control to quality assurance and TQM

Mass production systems such as the one that Henry Ford had instituted at the River Rouge plant certainly produced the desired result if units of production were the only measure. Crucially, the cars produced were cheap enough for most Americans to afford them. Unfortunately, as managers at factories like Ford were eventually to discover, the use of mass-production systems, because they rely on the need to keep producing high numbers without considering the wasted effort, are not only expensive, but ultimately, they alienate what Deming called the most important part of the production line: the customer.

Mass production was originally proposed by Frederick Winslow Taylor in his book *The Principles of Scientific Management* (Taylor, 1911). Essentially, mass production requires simplification and standardisation. Thus, any person can, after some basic training, carry out any one step in the production process on a repetitive basis. However, in order to ensure that the end-product is actually correct in every way, a specialist inspector must be employed to carry out checks. This person will have specific responsibility for finding faults – a task unlikely to endear him or her to the people who actually carry out the production. In order to remedy problems that they discover, inspectors frequently use what are called quality control techniques to *force* the workers to adopt practices that will stop the recurrence of such faults. As those who have used mass-production techniques usually find in practice, the desire to rigorously enforce inspection and control methods not only alienates the workers but also it is usually unsuccessful in detecting faults before the customer receives the product. Thus, despite the cost of finding and fixing problems which may occur, the customer still receives something that is highly likely to fail in practice. As anyone who buys something that doesn't work properly will tell you, this causes considerable annoyance.

In order to attempt to allow workers to exercise greater control over the task they carry out, quality assurance was developed. Thus, rather than assuming the inspector will find faults in an item subsequent to its production, a system is implemented whereby the worker is made responsible for checking the quality of their own work. Thus, in order for the worker to show that they have indeed checked their own work, it is necessary for them to produce documented procedures (these will allow someone else to see that what is supposed to happen is indeed occurring). Such a system, usually referred to as a quality system, such as ISO 9000 (BSI, 1994), can provide the basis for considering methods to produce long-term improvement (a point that will be considered in more detail in Chapter 5). On the other hand, a criticism that is often levelled at the use of quality systems is that once they have been created, there is a reluctance to change them very often; the effort that is required to write a quality manual makes this a daunting task. The result, therefore, is that unless someone can find a good reason for not adhering to the procedure which governs his/her task, they are expected to comply – regardless of whether or not the procedure actually ensures the customer gets what they want. In theory, the latter is remedied simply by rewriting the procedure(s): in practice, it is easier to do 'what the procedure in the book says'. As British managers in the mid-to late 1980s found after attempting to use QA to produce improvement, their counterparts in the USA were coming to terms with the fact that the apparent success of Japanese industry had been based on simple concepts taught to them by Americans in the 1950s. Moreover, as they learned from considering the development of what was usually referred to as TQM, the most important thing to be done in order to improve was to allow the workers a greater say in how the production processes should be carried out to achieve customer satisfaction. The development of this thinking is summarised in Fig. 3.5.

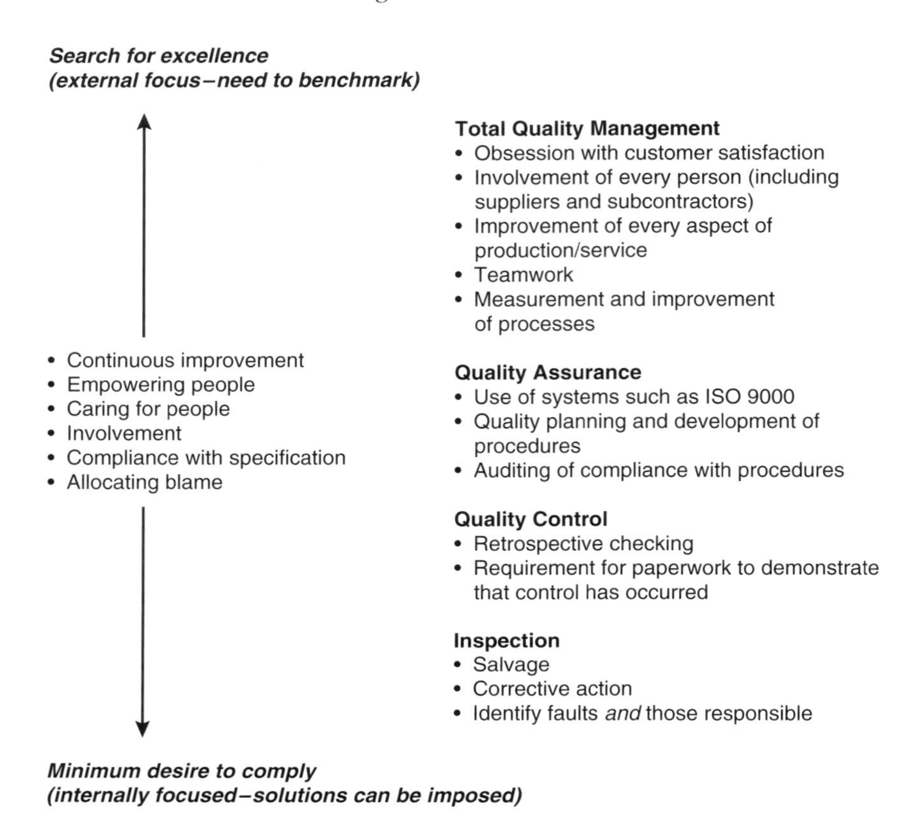

Fig. 3.5 Development of Quality Management.

3.5 *Achieving customer delight – the importance of recognising people as a key component of TQM*

Customers and, crucially, their perceptions about what they receive, are the key objective of TQM. Achieving customer delight (advocates of TQM and excellence argue that satisfaction is too limited (see below), is not an easy task; it will need considerable effort by every person involved in the supply chain. However, if there is any doubt about the importance of the customers, and of people in meeting their needs, it is worth remembering that those organisations that have found difficulty in trading in recent years usually discover that what they provide is no longer considered to be value for money. As such organisations also 'suddenly' find, there are others capable of doing it better or cheaper, and not infrequently, both. Analysis of how it is possible for organisations to be able to effectively 'steal' customers, strongly indicates that there is a combination of two things:

(1) Knowing how to refine the processes by which the product or service is created
(2) A focus on the use of people by which these processes can be operationalised and continuously improved

The central principle of TQM is that the process of improvement begins with the customer and 'ends'[17] with the customer. If an organisation does not know what its customers really want to receive, how can it ensure it meets their expectations? More crucially, if the organisation is unclear about its customer's expectations, how can it ensure that its employees are working in such a way as to ensure the attainment of such expectations?

Deming believed that the most important measure of an organisation to perform was not just its ability to achieve customer satisfaction, but *customer delight*. The former, he explained, is passive and suggests an intention merely to 'do enough'. As this book will stress, organisations that achieve the status of being regarded as 'world class', do so on the basis of consciously aiming to go beyond satisfaction. Such organisations are notable because of the apparent determination by employees, regardless of hierarchical level, to pursue proactive strategies in order to create value-adding at every stage. The ultimate objective of doing this is to create delight for customers at every contact they make with the organisation. As a test, the next time you contact an organisation, consider the way you were treated. For instance, how did the person who answered the telephone treat you? If you visited the premises of this organisation, was the way in which you were dealt with by its staff – from reception onwards – courteous, and such as to indicate that they valued your custom?

The treatment a customer receives, particularly at the first point of contact, is something that Jan Carlzon of Scandinavian Airways Systems (SAS) called the 'moments of truth'. Carlzon was appointed president of SAS in 1981 when it was making an annual operating loss of $20 million. Clearly this was an organisation in great difficulty. Amazingly, within a year, Carlzon had turned this loss into an operating profit of $54 million by the concept of 'moments of truth'. As Carlzon explains:

> Last year, each of our 10 million customers came into contact with approximately five SAS employees and contact lasted an average

> of 15 seconds each time. Thus, SAS is 'created', in the minds of our customers 50 million times each year, 15 seconds at a time. These 50 million 'moments of truth' are the moments that ultimately determine whether SAS will succeed or fail as a company. They are the moments when we must prove to our customers that SAS is the best alternative. (Carlzon, 1987: p. 3)

Thus, as Carlzon describes, he had to ensure that he and his managers created a system in which all employees who dealt with potential or actual customers were able to provide the best possible 'moments of truth'. As Carlzon maintained to his managers, staff in the front line were the ambassadors of SAS and if they felt the organisation did not value them, they would hardly feel inclined to put all their effort into pleasing the customer. Therefore, he reasoned these people had to have confidence in themselves. This involved SAS in addressing not just the immediate issue of training and procedures, but also the conditions in which their employees worked. Notably, one of the things that SAS did was to provide uniforms that were designed to the highest quality by Calvin Klein. As a consequence, SAS employees were enabled, *empowered* (a word that has great significance in improvement initiatives such as TQM) to use their own initiative on matters that were previously only capable of resolution by their superiors; something which inevitably took time and caused annoyance to customers.

According to Bank, what occurred at SAS was similar to the experience of Rank Xerox (Bank, 1992: p. 116). In particular, he describes how SAS was transformed from an organisation which had a 'technical production orientated attitude' to one in which there was an 'almost evangelistic dedication to putting customer service above all else' (Bank, 1992: p. 19). This transformation, he explains, occurred because employees in both Rank Xerox and SAS were encouraged to operate in a very different way to that which occurred previously. Additionally, Bank argues, there was an obsession in the desire by every person in these organisations to continuously think about how their approach to customers could be improved. What Bank is alluding to is something that is more commonly referred to as the culture of an organisation.[18] As the next chapter describes, a vital part of benchmarking and improvement is the need to consider the culture of an organisation. However, as will be explained, dealing with culture is something that requires great skill, understanding and, usually, considerable time and effort.

Summary

In this chapter, the following aspects of TQM have been described:

- What TQM is, and why it is important to benchmarking
- Its historical development in Japan from the philosophy of SPC (Deming) and Juran's quality trilogy
- How Toyota learned to understand the value of process management and employee contribution to improvement
- The use of TQM in the West, and how it is possible to learn (benchmark) how excellent organisations have achieved superior levels of customer satisfaction

CHAPTER FOUR
FACILITATING A CHANGE IN ORGANISATIONAL CULTURE

Objectives

In the previous chapter the concept of TQM was described. This chapter explains why creating a change in organisational culture is a vital part of developing TQM. Specifically, this chapter explains the following:

- What the issue of organisational culture is
- Approaches to how the culture of an organisation is addressed in order to produce change
- The role of change agents in producing improvement using benchmarking
- Why encouraging learning by all employees is crucial in producing improvement

4.1 Organisational culture

Anyone who has been interested in the subject of quality management over the last ten years will probably have been referred to a book called *In Search of Excellence* which was written by Tom Peters and Robert Waterman and published in 1982. This book has become notable for that fact that it is the highest selling management text of all time (over five million copies to date). The reason why this book should have proved so popular has been the subject of debate in recent years. As Crainer suggests, it is possibly because of a perception that it 'accentuated the positive at a time of unmitigated gloom' (Crainer, 1996: p. 113). As he suggests, managers welcomed the message that by following advice that Peters and Waterman had derived from studying 62 'excellent' companies in America (particularly in respect to what they called the 'eight attributes of excellence'), they too could emulate the apparent success that these companies enjoyed. It is worth pointing out that *In Search of Excellence* has, despite its sales record, been criticised for having over-simplified what is required to become excellent. Moreover, as many commentators on quality management argue, if the eight attributes that Peters and Waterman provided were so effective, why have so many of the companies they cite as being excellent ceased to exist due to bankruptcy?

The reason for specifically mentioning Peters and Waterman's book is that, despite the criticism it has received, it drew attention to the fact that in order to become excellent, it is necessary to address the issue of organisational culture. As they contend:

> Without exception, the dominance and coherence of culture proved to be an essential quality of the excellent companies. Moreover, the stronger the culture and the more it was directed toward the marketplace, the less need there was for policy manuals, organisation charts, or detailed procedures and rules. (Peters & Waterman, 1982: p. 75)

Thus, managers, having read *In Search of Excellence*, were left in no doubt that if they wished to create excellence in their organisations, it was essential to consider how to address organisational culture.

4.1.1 What is organisational culture?

Perhaps the biggest problem that 'culture' suffers from is the lack of an agreed definition – there are literally hundreds to choose from. Whilst many of these definitions appear to be very similar, each provides a slightly different interpretation of an organisational issue which, despite seeming at first to be relatively simple, in practice involves an understanding of things which determine how people behave. Jaques is one who provides an explanation that encapsulates all the major elements of what culture involves:

> The culture of an [organisation] is its customary and traditional way of thinking and of doing things, which is shared to a greater or lesser degree by all its members, and which new members must learn, and at least partially accept, in order to be accepted ... [it] covers a wide range of behaviour: the methods of production; job skills and technical knowledge; attitudes towards discipline and punishment; the customs and habits of managerial

> behaviour; the objectives of the concern; its ways of doing business; the methods of payment; the values placed on different types of work ... and the less conscious conventions and taboos. (Jaques, 1952: p. 251)

What this fairly long definition describes is what might be regarded as the 'social glue' that holds people together. Many commentators believe, similarly to Peters and Waterman, that the stronger the culture which exists in an organisation, the more that people will adhere to the values and accepted practices that are believed to exist. It is perhaps no surprise that very strong cultures usually exist in informal organisations such as religious groups; the values that exist will have been put in place over many generations, and will only be changed for special reasons. Thus, as Peters and Waterman assert in *In Search of Excellence*, if the management of an organisation wish to ensure that its customers receive the best service possible, it is essential that all employees believe in this ethos. As those who studied Japanese organisations discovered (by what is now called benchmarking), there was a very strong commitment by every employee – including suppliers – to adhere to the principle of giving the customer the 'best'. The challenge is in what needs to be done in order to achieve change in the culture of the organisation which will create this sort of cohesiveness among the workforce.

4.2 *Senior management's role in creating cultural change*

Chapter 3 cited two examples of organisations in which change had occurred that resulted in the creation of extremely strong commitment to customer service (Rank Xerox and SAS airways). One thing that seems clear in both these cases is that the effort to change was led by an individual with an extremely strong sense of vision and purpose. Whilst the efforts of those at lower levels will be crucial in creating what Jan Carlzon at SAS called the 'moments of truth', it is undoubtedly the case that unless the management team at the most senior level are absolutely committed to achieving change, any initiative will fail. As will be described in Chapter 7, one of the criteria for the EFQM Business Excellence model is leadership. Thus, senior managers in any organisation using benchmarking as a way to improve must be prepared to do the following:

- Demonstrate their absolute support for continuous improvement
- Stress the need for value as opposed to cost

- Ensure that the philosophy of giving the customer the best possible is articulated at every opportunity
- Institute methods of communication that enable people to provide their opinions about what needs to be done
- Ensure that employees are aware of the need to address customer–supplier relationships
- Support training and education of all employees to ensure that every person can perform their part in the process of improvement
- Ensure that the management systems that exist are optimal for facilitating continuous improvement (remember Deming's belief about management responsibility for creating management systems which reduce variation)
- Institute an atmosphere where people are encouraged to co-operate
- Encourage effort, and never to blame if failure occurs (as will be described later in this chapter, a vital part of the process of benchmarking and improvement is the willingness to learn from mistakes)
- Implement strategies which support and integrate all of the above

4.2.1 What senior managers in construction organisations can do to create culture change

The list shown above is partly based on the advice of Juran (1993). As he recommends, it should be considered to be the minimum necessary. Consultation of texts which describe TQM/improvement/benchmarking indicated that the issue of senior management commitment can be considered to be the most fundamental component. As writers of such texts assert, if an organisation is to achieve a shift in the emphasis on customer service, it cannot happen without the active participation of managers at the highest level. What this means is that managers may have to totally reconsider how they approach organisational matters such as customers and people (see below). However, the culture that exists in an organisation will usually be the result of many things, not least its historical development. As Fig. 4.1 below shows, the culture of an organisation is often the result of being perpetuated by those who believe it suits their objectives. The reason for this is probably obvious. In the past, most senior managers were appointed because they were trusted to continue the ways (customs) that had been

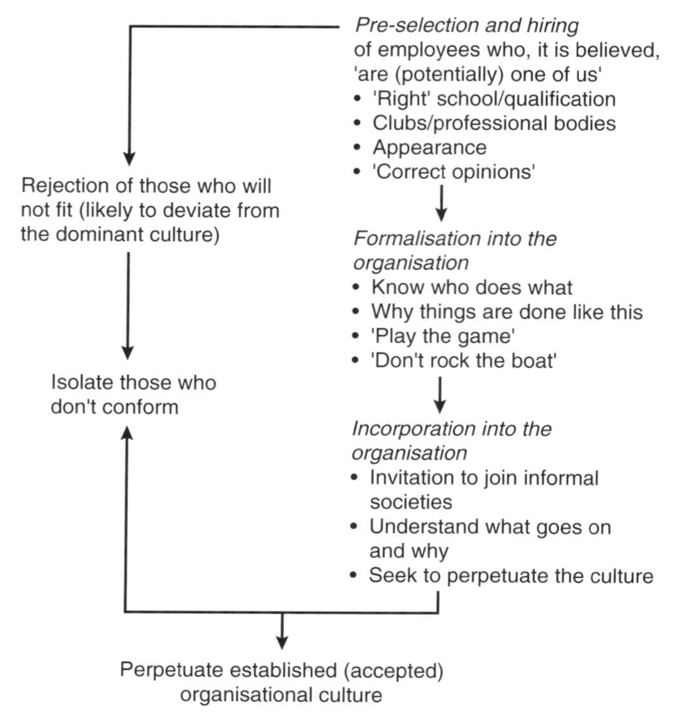

Fig. 4.1 Ensuring continuance of the established culture.

established previously, and importantly, were believed to have worked perfectly well.

Construction is not atypical in its tradition of appointing those who were expert in particular technical areas. As a consequence, there has been what might be interpreted as an obsession with ensuring that targets – most particularly those involving cost – are met. The result of this obsession is an industry in which disputes arise between clients and contractors – what many describe as the 'adversarial nature' of the construction industry. As many studies into construction have shown (most notably that by Latham, 1994), the culture of the construction industry is one in which clients are regarded more as enemies than valued customers. The challenge, therefore, is for senior managers in construction to commit themselves to dedication to improvement in a way that is absolutely explicit. For instance, the senior management of Ford Motor Company, in their annual vision and mission statements, state that providing their customers with cars of the highest performance is the number one objective.

Therefore, as part of a process of improvement, senior managers in construction should, in order to act as *role models*, do the following:

- Visit excellent organisations to see how their counterparts carry out their duties
- Talk to clients on a more regular basis to find out what they really want from your organisation
- Using typical Tom Peter's vernacular, carry out 'management by walking about', i.e. go out more regularly and talk to staff who carry out day-to-day operations
- Become actively involved in training and educating staff in how to use quality tools to produce improvement
- Be prepared to encourage staff to openly discuss problems/ deficiencies that currently exist in the organisation
- Set up forums and social events to celebrate achievement of success
- Constantly ask the question, 'How can I, as part of the senior management team, do more to assist my staff to achieve their best?'

This is not an exhaustive list. However, it is absolutely essential to stress the need for improvement to be seen as something that is never-ending. Thus, senior managers must never allow themselves to believe that once they have done 'enough to get things going', they can leave it to others. Just as destructive, would be the decision, because of limited budgets, to decide that after a certain period, improvement will end. Dale and Cooper, in stressing this point, argue that it is this assumption by senior management who treat TQM as a 'programme', which causes their organisations to be 'overtaken' by competitors who treat it as a continuous process (Dale & Cooper, 1992: p. 14).

4.3 The role of middle managers and change agents

Even though the role of senior management is one that is stressed by all advocates of cultural change, there is a caveat. Whatever senior managers are able to do, there is a limit to their ability to produce the actual change in people's attitudes and behaviour that is crucial. Senior managers are, after all, the custodians of the organisation, and despite the importance that is attached to their active involvement in encouraging others to support efforts to improve, they will have other duties. Senior managers, regardless of how much energy they possess, cannot be everywhere at once.

Because of the inability of senior managers to be able to be involved in the day-to-day culture change, it is necessary to appoint less senior managers who will act as co-ordinators for all the activities that carrying out improvement will involve. This, it must be stressed, is advice that much of the literature dealing with culture change tends to ignore. There is, it appears, an overemphasis on 'selling the idea' to those who will make the strategic decision, and little attention to the task of those who must operationalise this intention. As Kotter and Heskett (1992: p. 93) explain, middle managers (e.g. departmental heads), even though they cannot initiate culture change, tend to be those who will have most influence on the staff who carry out day-to-day activities. In order to do this they should be prepared to carry out the following duties:

- Develop plans for their own departments which are consistent with the strategic plan for the whole organisation
- Assist in the training and education of their staff who will be using improvement tools; this will require them to act as a coach or mentor, *not* as their line manager
- Communicate why it is vital to engage in activities which will focus on enhancing customer satisfaction
- Provide feedback to senior managers as to any additional resources that are required
- Carry out measurement and benchmarking to monitor the effects of improvement

Whilst the role of middle managers in culture change receives little attention, there is another middle manager who is usually appointed to the role of central co-ordinator for all improvement activities. In effect, this person becomes the '*organisational champion*' for change. Research carried out for a previous study was entirely focused on this sort of manager (McCabe, 1999). What was discovered during the course of a study of 12 quality managers[19] working for construction organisations was that all had been appointed to implement QA. However, as was found by exploring the working lives of these managers (see McCabe, 1997) a number of these managers did more than just manage the quality system. Specifically, these managers had either been instructed to develop change based on the principles of TQM, or had come to the conclusion that 'merely doing QA' would not be enough to cause a radical shift in the way that activities were carried out in their organisations.

These managers found that they became the one person who coordinated all activities associated with quality management (including, it should be noted, benchmarking). As these managers explained, even when the decision to 'move beyond QA' had come from their superiors, it was incumbent on the managers themselves to monitor the performance of senior management. In some cases, this involved them in putting pressure on those above to demonstrate to employees their continuing commitment to the quality initiative. This task, they admitted, caused some discomfort; it takes great confidence in one's ability to be able to tell one's bosses that unless they do more, the improvement initiative will fail. As a consequence of this finding, it is essential that the persons appointed to deal with initiatives for culture change are chosen with an emphasis on their ability to inspire people to be willing to attempt alternative methods of carrying out their functions. Most particularly, there are four elements that should be considered when making the appointment of someone who is capable of 'driving cultural change' through the various organisational levels:

- Leadership skills
- Motivational skills
- Skill in dealing with resistance
- Skill in recognising different approaches, values and norms (culture)

The sort of person appointed to manage the implementation of culture change and TQM will need to be someone who, as well as being absolutely competent in the technical issues surrounding the subject, possesses abundant confidence and human relations skills. As this person will frequently find, they will be expected to support, cajole and defend the need to engage in change. As the author discovered during his research, whilst resistance may come from below, it can also come from above; senior managers fear that trusting their workers (empowering them) is like 'letting go of the controls'. Few people want to tell those above them that they are part of the problem and need to demonstrate their willingness to engage in change. However, the real 'acid test' of any organisational culture change is being able to win the hearts and minds of those at the sharp end. This is dealt with in the next section.

4.4 *'Getting the troops on board'*

No matter how committed the senior managers and middle managers of an organisation are to the culture change initiative, there comes the problem of how to ensure that those whose day-to-day lives will be most affected are willing to support it. As employees below senior and middle management level normally constitute the majority of those who belong to an organisation, it is what these people really believe and value that will create the predominant culture. As a result, if the change initiative is not willingly supported by people at this level (because, for example, it has been imposed by the use of threats), its continuance will be entirely dependent on the need for managers to monitor their staff to ensure compliance with the 'new way of thinking'. This is not the sort of approach that Deming had in mind when he advised Japanese organisations to ensure workers have 'pride in their job'.[20] Moreover, the sort of change that requires constant threats will hardly achieve satisfied and dedicated workers. Therefore, it is vital that those who are responsible for introducing change, do so in a way that is sensitive to the value systems of those who will be directly involved in the process of delivering the product or service to the customer. It was for this reason that SAS provided uniforms that were of a quality that reinforced the message that its employees were essential in giving superior customer service. As will be described in more detail in Chapter 6, which deals with improving customer satisfaction, the 'people' aspect of any organisation is something that requires a great deal of attention.[21]

Inevitably in any process of change, there will be some resistance. Not all people welcome change, particularly if it involves them in using new methods or routines, whatever the merits of doing so. However, if people are aware of why change is required, and more particularly, have assisted in the development of the changes, the likelihood of resistance will be reduced. Even so, it will still be vital for those who are responsible for the introduction of change to be aware of any potential concerns that people may have, especially if retraining or moves to new departments are required.

The issue of why people should want to change needs to be addressed at the earliest opportunity. This can take many forms:

- Workshops
- Discussion groups
- Teambuilding exercises (see section 4.4.2 below)
- Social events

What has been found to be important in attempting to create improvement is that employees derive satisfaction from what they are doing. This introduces the aspect of employee motivation.

4.4.1 Motivation of people

Motivation is, it must be stressed, a vast subject in its own right. However, a useful working definition would consider the 'forces acting on or within an individual which initiate and direct behaviour' (Sims *et al.*, 1993: p. 273). As we are all aware from our own experiences, there are forces that are self imposed; we do it because we feel a sense of duty or honour. These forces, often referred to as intrinsic, are usually powerful. Unlike the other type of force – that which comes from external sources (extrinsic) – intrinsic motivation does not require managers to make employees do things. So, for instance, in an organisation striving to become more customer-focused, the aim will be to encourage its workers to believe that it is their duty to do everything possible to give the customer[22] the best service or product possible.

Motivation theories have existed since the 'discovery' by researchers at the Hawthorne plant (the location referred to in Chapter 3 as being notable for having employed Deming and Juran) that the satisfaction that people derive from their jobs frequently depends on more than the external conditions imposed by managers. As such, there was a realisation that consulting workers about how best *they* think that the task they carry out should be achieved can increase intrinsic satisfaction. Therefore, managers, acting in accordance with the principles that Deming espoused, have a duty to create a system in which employees of an organisation feel that they are both encouraged, and recognised for giving extra effort. There are many things that should therefore be considered:

- What skills do our employees currently have?
- What additional training and education do our employees need to be able to contribute to the improvement process?
- How can we encourage people in this organisation to more actively consider methods of carrying out day-to-day tasks that will create opportunities for improvement?

A word that is frequently associated with this approach to management is *empowerment*, defined by Sims *et al.*, as follows:

> ... the idea that, given the freedom, scope and resources to achieve organizational *goals*, people will, in effect, lead themselves – if it is in their interests to do so. (Sims *et al.*, 1993: p. 246)

In many organisations, it is quite probable that employees have been used to simply doing what they are told to do; being asked what they think can be done may be an alien concept. This is something that many managers, particularly at middle management level, may also find difficult. As Sims *et al.* suggest, empowerment will result in a 'democratisation' of the organisation, the result being that managers are no longer expected to be seen as authoritarian figures – something that these managers often feel is a diminution of their power.

Sirkin (1993: p. 58) believes that in order to empower employees, there are five essential features that should exist:

(1) Managers need to understand what it is like to do the job; they must have *empathy* with what workers are expected to achieve
(2) Those who are to be empowered must be given appropriate responsibility for their task (they must also be allowed to make mistakes)
(3) Empowered workers need to be given training which is at least adequate to achieve the desired outcomes
(4) In addition to (3), it is essential that those who will be empowered are provided with the resources that are necessary
(5) As Sims *et al.* point out with respect to democratisation, even though the decision to use empowerment will have been taken at a senior management level, it is those at middle management level who are most likely to feel the loss of power. Accordingly, these middle managers will need to be convinced (educated and trained) to realise that the use of this technique will benefit the organisation as a whole.

This book has referred to the many theories that exist with respect to motivation and it is not the intention to deal with all of these (most management texts do this adequately (see in particular, McCabe, 1998)). However, there are two theories that it would be useful to explain which capture aspects of motivation of people that are sympathetic to TQM improvement, and therefore, assist in producing culture change. This first, is one that is proposed by Hackman and Oldman (1980). Essentially, it draws attention to five factors – *core job characteristics* – that result in *three critical psychological states*, and which result in four improved *personal and work outcomes*. Additionally, there are what Hackman and Oldman call '*moderators*' which are largely determined by an individual's ability to develop themselves. As Hackman and Oldham propose, managers, by understanding the implications of this model, can

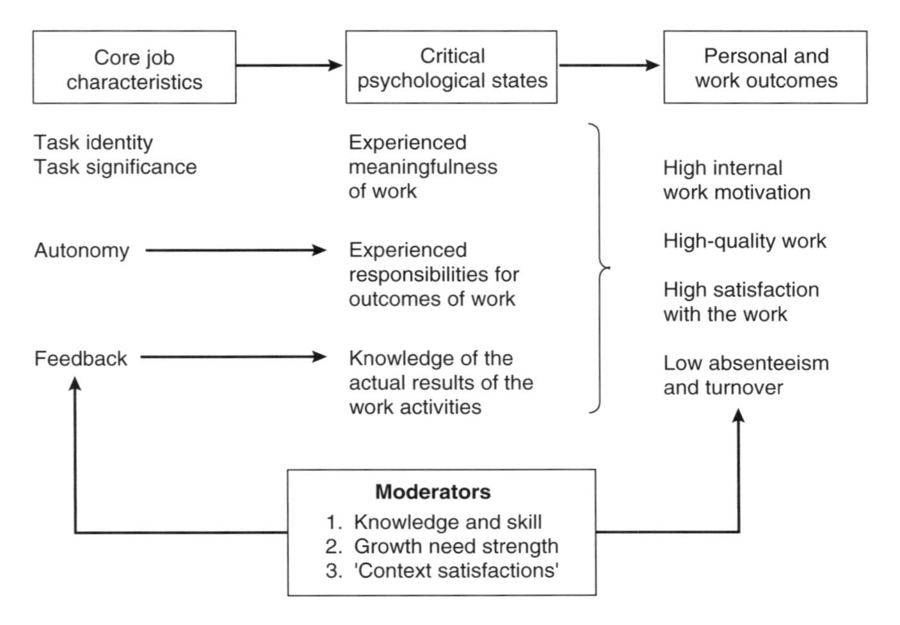

Fig. 4.2 The Hackman and Oldman job characteristics model of motivation.

assist individual[23] workers to derive greater satisfaction from what they do. This model is displayed in Fig. 4.2.

The second theory is one that links how managers can assist employees to increase their ability to attain personal and organisational goals. This model, 'The Path–Goal Theory of Leadership' (Bernie Bass, 1985), shown in Fig. 4.3, suggests that it is important for managers to explain how and why it is essential for employees to attain the organisation's goals of increased customer satisfaction.

Whilst motivation theory is essentially about individuals, there is a recognition that the satisfaction that an individual derives is influenced by the relationships they have with others with whom they have to work.[24] Thus, implicit within the desire to create an organisation in which all employees are motivated towards the goal of organisational improvement, is the need to facilitate teamworking. What this involves, and how it may be achieved, is described in the next section.

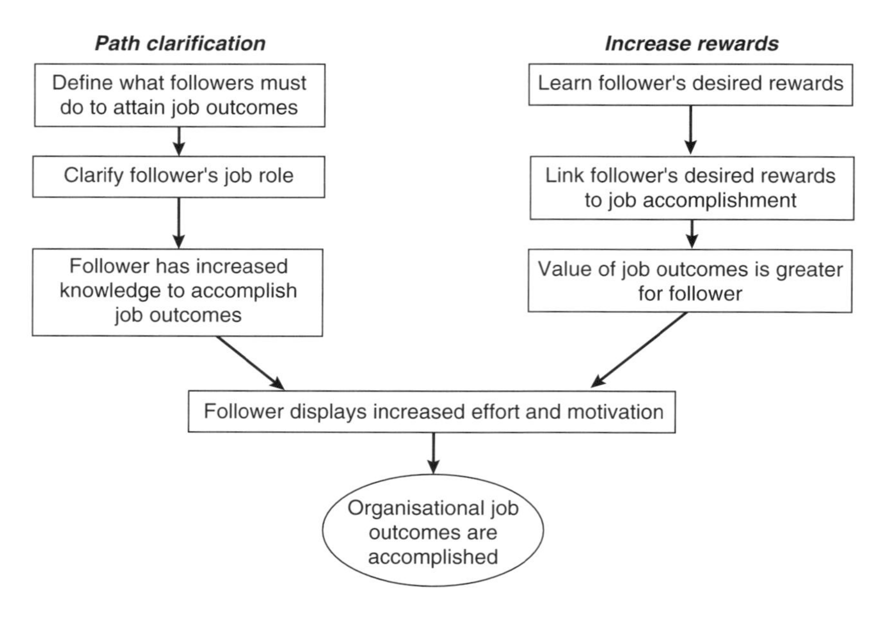

Fig. 4.3 The path–goal theory model.

4.4.2 Using teamwork in cultural change

John Oakland (one of the most influential British writers on TQM), in his book *Total Organizational Excellence – achieving world-class performance*, believes that an essential part of organisational improvement is the need to encourage teamworking (Oakland, 1999: p. 17). As he explains:

> Barriers are often created by 'silo management', in which departments are treated like containers which are separate from one another. The customers are not interested in departments – they stand on the outside of the organisation and see slices of it – they see the *processes*. It is necessary to build teams and improve communications around the processes. (Oakland, 1999)

Oakland provides the following reasons why the use of teams will prove to be superior to individuals working on their own:

- A greater number of issues can be dealt with
- Problems that are too complex for one person to cope with can be tackled
- There will be a sharing of expertise and knowledge

- The fact that others share their concerns will mean that the perception of isolation will be reduced
- Co-ordination of effort will result in anticipation of potential problems
- Conflicts which arise between departments and inter-organisational groups as a result of misunderstanding or interpretation of requirements are likely to be reduced
- Implementation of solutions to problems will have consensus; the chances of success are therefore increased

In addition to this list, Dale and Boaden suggest that as a result of using teams, the following things will occur (Dale & Boaden, 1994: p. 515):

- People will be more committed to the principle of TQM
- It is likely that the needs of customers will be understood
- There will be greater understanding between managers and workers (who co-operate in teams) about the decisions taken and their impact on the business
- Morale will increase among employees if success is achieved in organisational goals
- Improvement will become an accepted part of the culture of the organisation

As you may have heard, the word team can be considered to be an acronym for 'Together Each Achieves More'. Teamwork, therefore, is an essential part of TQM and improvement. Indeed, as Stahl, using examples of organisations such as Motorola and the Florida Power and Light Company[25] argues, teams that focused on 'cross-functional issues' were the 'key' to transforming these organisations and the managerial practices which existed within them (Stahl, 1995: p. 387).

4.4.3 Types of team

The method of choosing who should become part of the team is discussed below. However, there are various types of team that people could form themselves into:

(1) *Cross-functional teams* usually involve members from different departments that interrelate with one another. For instance, on a construction site, when a part of the building is being

planned, it is advisable that the major trades which will contribute should form a team to discuss how best to co-ordinate activities. By doing so, it is likely that innovative solutions to the scheduling of construction may emerge. More importantly, the fact that representatives from different trades communicate with one another will enhance understanding of each other's problems; it is the failure to see it from 'the other side' that has led to disputes arising in the past.

(2) *Project teams* are similar to cross-functional teams in that they exist in order to achieve a specific outcome, i.e. the construction of a finished building. The difference, therefore, is that this team will usually consist of representatives of the main parties with an interest in the project (client, designer, builder, quantity surveyor, etc.)

The existence of these teams is fairly normal in construction. However, they do have the following characteristics that possibly undermine the ability of employees to create improvement opportunities:

- The objectives may have been set by those external to the group
- The team is led by someone on the basis of their seniority
- They can be temporary in order to address a specific problem or objective
- Members may be chosen randomly without consideration of the skills or expertise that individuals possess (people may be instructed to attend by their line manager)
- Meetings may sometimes be held on the basis of being *regular* (for instance weekly), rather than when necessary, resulting in the activity becoming resented

As a consequence, and whilst it is stressed that cross-functional and project teams are likely to produce some excellent results, it is the following two types of team that are most widely associated with improvement and TQM:

(1) Quality circles
(2) Improvement teams

Quality circles

This form of team is, according to Dale, 'a natural part of Japanese working life' (Dale, 1994: p. 104). In effect, quality circles exist in every section/department/office and consist of workers and managers involved in carrying out a particular function. The objective is that every person in the team is constantly trying to think about how the team might carry out the task in such a way as to improve the end result.

A quality circle is quite explicitly a form of team where membership is voluntary. The philosophy of a quality circle is that between six and eight members (it is advisable not to exceed the limit to avoid difficulties with group dynamics) meet on a basis of attempting to both problem-solve, and to create solutions for improvement. The rules that govern the operation of the quality circle are as follows:

- Members are of equal status; if a chair is required he/she will be elected by the group (some advocates specifically advise that this person should not be the most senior)
- Members of the group are free to choose any problem that affects their task(s) within their section/department/office
- The emphasis is on creating solutions which the members can implement themselves
- The members will have been trained in the use of using simple statistical methods, meeting skills, facilitating actions, measurement and presentational techniques
- Meetings should be short but frequent (possibly every morning for, say, half an hour)
- If it is necessary to hold meetings out of hours (because of the need to keep the section/department/office operational), members will be paid
- If requested by the group, an external facilitator will be provided[26]
- If additional resources are required in order to implement solutions, the group will do so by delegating the chair to present this request formally
- The members will constantly evaluate the effects of the solution in order to determine the rate of improvement, and therefore be better able to judge the potential of future solutions to different problems

Improvement teams

This sort of team is, like quality circles, aiming to create solutions to problems and attempt to create, as the title makes clear, improvement. However, the main difference is that whereas the quality circle consists of members of a section/department/office, this team exists to consider the potential for improvement in areas such as waste, customer satisfaction and productivity across the organisation. In addition, the team will try to include members from organisations that have a direct interest in the final product or service that is being created, i.e. customers, suppliers and subcontractors. As such, this group may also appear similar to the cross-functional group described above.

The main features of an improvement team are as follows:

- Its membership is ideally voluntary although some may be mandated to attend
- The focus is usually to consider particular problems that have been discovered by using customer surveys or employee suggestions
- As a result of the need to create action, it will be usual for member(s) of senior management to attend (even though, like the quality circle, this person(s) will not automatically be chosen to lead the group)
- Like quality circles, members of this team will use problem-solving techniques, simple statistical methods, and monitoring of the effects of potential solutions (this may require training)
- Solutions that require additional resources or changes to existing processes will be presented to senior management for agreement
- In order that potential solutions are implemented successfully, it is essential that those who are affected, but who do not belong to the improvement team are briefed as to what is happening and why

Overview

Manson and Dale (1989) provide a comparison of quality circles to improvement teams. This comparison is summarised below in Table 4.1.

4.4.4 Picking the right members for a successful team

In any team, success in attaining objectives is usually achieved by bringing together the right mix of talents. Whilst it might seem desirable to construct a team of people who, because of their shared interest, are likely to get on well, this is not normally a primary consideration. However, there is a view that having people who are

Table 4.1 Quality circles compared to improvement teams

Characteristics	Quality circles	Improvement teams
Purpose	Involvement of people Increase participation Develop culture of improvement	Improve general aspects of the business Consider problems which will enhance customer satisfaction
Team-building	Members from certain department Effective understanding of nature of problems Consensus of solution	Formed around problem-specific task Team develops according to requirement placed upon it by originators of problem
Leadership	By consensus (voted) Members have power confined to group Dependent on others for assistance to provide extra resources	Managers/supervisors from 'interested' sections or other organisations Independent
Problem-solving potential	Limited Minor problems which concern their task(s) Limited skills	Considerable Major problems Highly skilled
Project resolution rate	Low	High
Infrastructure	Informal 6–8 members Report to senior managers when necessary	Formal Report to senior management Steering committee

too similar will not, ultimately, be constructive. It is suggested that one of two things may occur. Firstly, because these people have such similar views, there will be no disagreement about any aspect of what they do. Alternatively, precisely because these people are so similar, there will be no agreement about who should do what.

It was the issue of how teams should be constructed that concerned the psychologist Dr Meredith Belbin. Based on empirical research, Belbin observed that teams which worked successfully together were not made up of people who had the same abilities or interests. Instead, he discovered, there was a range of roles which the team members performed. In his book *Management Teams why they succeed or fail* (1981), Belbin contends that there are eight roles that are required for any group to interact together in order to facilitate the attainment of its aims or objectives. These eight team roles are summarised in Table 4.2.

4.4.5 Development of the team

Oakland asserts that any group of people who come together to complete a task or series of activities must undergo a transition from being individuals who are independent of one another to a team which is interdependent (Oakland, 1999: p. 157). As he explains, there are three phases – each of which is a precursor to the next – that a group must undergo in order to become an effective team:

(1) *Phase one: 'exchange of basic information and ideas'* during which members become used to discussing their beliefs about how things might be done differently.
(2) *Phase two: 'trust'* during which members become accustomed to sharing their ideas. As Oakland believes, it is at this point that 'the elimination of fear' (Oakland, 1999) occurs.
(3) *Phase three: 'free communication'*: it is at this point that members can operate in a way which is interdependent, and which is, according to Oakland, 'critical for continuing improvement and real problem solving' (Oakland, 1999).

Tuckman and Jensen (1977) propose that in team development there are four distinct stages. Whilst these phases are not different to the three phases that Oakland provides, what they indicate is that team development is a process that involves some difficulty.

(1) *Stage one: Forming – awareness*
During this stage, people become aware of one another and, in particular, understand each other's aspirations. Tuckman and Jensen advise that it is important that people in the team are aware that unless this stage is accomplished properly, all subsequent stages will be compromised. In particular, it is important to be aware of the following:

    (a) That objectives are established so that everyone is clear about what they must achieve
    (b) Concomitant with the above, that everyone is clear about their role in the team (this may involve the use of techniques to decide who is best suited[27])

Table 4.2 Belbin's roles in teams

Type	Symbol	Typical feature	Positive qualities	Allowable weaknesses
Company worker	CW	Conservative, dutiful, predictable	Organising ability, practical common sense, hard-working self-disciplined	Lack of flexibility, unresponsiveness
Chairman	CH	Calm, self-confident, controlled	A capacity for treating and welcoming all potential contributors on their merits and without prejudice. A strong sense of objectives	No more than ordinary in terms of intellect or creative ability
Shaper	SH	Highly strung, outgoing, dynamic	Drive and readiness to challenge inertia, ineffectiveness, complacency or self-deception	Proneness to provocation irritation and impatience
Plant	PL	Individualistic, serious-minded, unorthodox	Genius, imagination, intellect, knowledge	Up in the clouds, inclined to disregard practical details or protocol
Resource investigator	RI	Extroverted, enthusiastic, curious, communicative	A capacity for contacting people and exploring anything new. An ability to respond to challenge	Liable to lose interest once initial fascination has passed
Monitor/ evaluator	ME	Sober, emotional, prudent	Judgement, discretion, hard-headedness	Lacks inspiration or the ability to motivate others
Teamworker	TW	Socially unemotional, rather mild, sensitive	An ability to respond to people and to situations, and to promote team spirit	Indecisiveness at moments of crisis
Completer/ finisher	CF	Painstaking, orderly, conscientious, anxious	A capacity for follow-through, perfectionism	A tendency to worry about small things, a reluctance to 'let go'

(Reprinted with permission from Belbin (1981, p. 78))

(c) Avoid attempting to replicate existing organisational hierarchical structures on the group (especially, they stress, the tendency to produce bureaucracy over action)

(2) *Stage two: Storming – conflict*
It is during this stage that team members may become embroiled in conflict. The reason for this, Tuckman and Jensen suggest, is that, like Oakland's explanation of team development, people are reluctant to relinquish their day-to-day roles (they cannot 'break out of themselves', something that may occur particularly among those in middle management). Like Oakland, Tuckman and Jensen advise that until people learn to trust each other as equal members of the group, progress to the next stage is impossible.

(3) *Stage three: Norming – co-operation*
During this stage, members of the team should begin to appreciate what each can contribute in order to achieve success. As such, they should have developed confidence in their own abilities and those of the other team members. As a consequence, the team will have started to operate in a much more systematic way towards the attainment of objectives that will have been previously agreed. This will require that the team agree the following:

(a) Plans of action
(b) Methods of collecting information
(c) Techniques for review of progress

(4) *Stage four: Performing – productivity*
By this stage, all the effort that will have been put into the previous stages should have resulted in a team that is capable of doing what it set out to achieve. Failure to do this, whilst not being inconceivable (due to unforeseen circumstances), may indicate that the team did not adequately agree this could realistically be achieved with the existing skills or resources. Most importantly, if the team is engaged in improvement activities, it should continuously measure the effects on processes of whatever actions it implements. As Tuckman and Jensen recommend, the group, if it is working well together, will be able to effectively respond to change. Moreover, and as will be explained in the next section, the group should be able to learn from what will have occurred, and which can be usefully applied to future situations.

Finally, as part of the explanation of how teams operate, Kormanski and Mozenter (1987) propose a diagram that summarises the stages and outcomes that result from teamwork in Fig. 4.4.

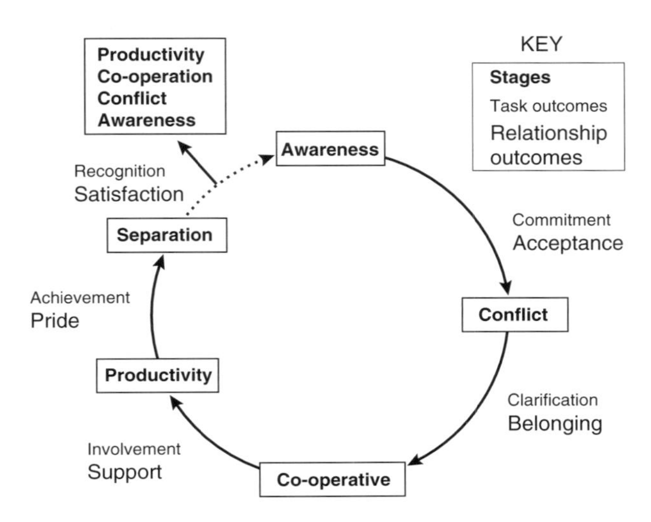

Fig. 4.4 Team stages and outcomes.

4.5 *The role of learning organisations in TQM and benchmarking*

Pedler *et al.* argue that unless an organisation is one in which learning by all its members is encouraged, it will be impossible to attempt to create sustainable transformation (Pedler *et al.*, 1991: p. 1). As many readers will already be aware, word 'learning'[28] is one that is now regarded as being crucial to organisational development and change. For instance, Burnes asserts that in order to create change, learning must become something that every employee is encouraged to do in order to effectively respond to ever-increasing customer expectations (Burnes, 1996: p. 191). It should be stressed that in order to carry out benchmarking, it is necessary to do more than simply copy best practice from another organisation. As those organisations which have been successful at achieving excellence have discovered, simply attempting to apply techniques that their competitors use will not result in radical transformation. What is required, therefore, is to be prepared to look at the processes used by another organisation which operates outside the sector and, having reflected upon the reasons for its success, to learn from this in order to consider how existing processes may be changed. It is for this reason that Karlof and Ostblom, in their book *Benchmarking: a signpost to excellence in quality and productivity*, deliberately draw attention to the need to engage in what they describe as 'benchlearning' (Karlof & Ostblom, 1993: p. 181). As they explain, what is vital in creating an excellent organisation is to create a cultural change among people which 'codif[ies] successful behaviour'. In Fig. 4.5 Karlof and Ostblom provide an excellent summary of the connection between *benchmarking* to achieve *efficiency* and *benchlearning* to achieve *proficiency*.

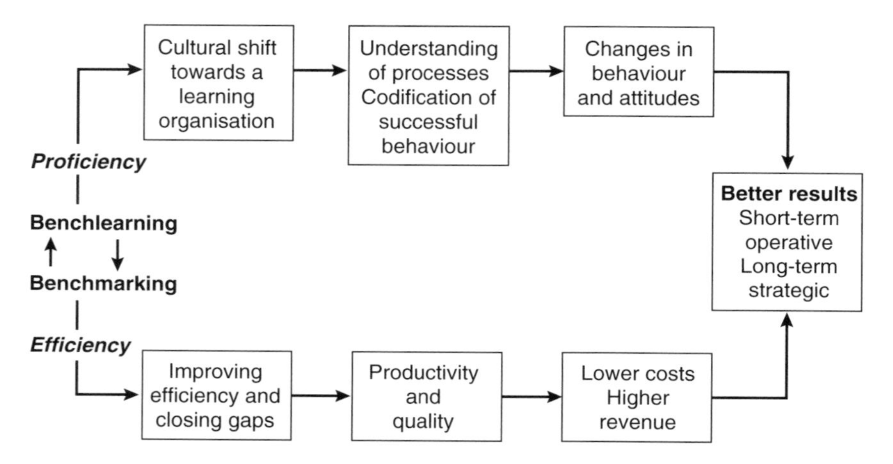

Fig. 4.5 Benchmarking and benchlearning as concurrent activities.

Mayo (1993) provides a number of characteristics that define a learning organisation. Some of these are:

(1) The use of the word *learning* becomes commonplace
(2) People's development assumes a high priority among managers
(3) Consultation of employees at operational level occurs as a matter of course
(4) Feedback from employees is encouraged
(5) The methods of learning that are used are decided upon by employees rather than managers (see next section)
(6) Mistakes or accidents are treated as opportunities to learn rather than being used as a means by which to blame and punish; remember that some of the greatest inventions that

have been discovered occurred as a result of accident rather than design
(7) Information is shared by everyone
(8) Openness and honesty among members is accepted as 'normal'
(9) Spontaneity and informality are encouraged
(10) The organisation continually seeks to benchmark itself against others in order to improve
(11) The use of learning is a method by which to make work seem like fun

4.6 *Methods of organisational learning*

A learning organisation, Karlof and Ostblom believe, is one in which employees are encouraged to 'constantly [seek] to acquire new knowledge and translate it into skills' (Karlof & Ostblom, 1993: p. 187). All advocates of learning in organisations argue that once people have become used to behaving in this way, their motivation to continue will become intrinsic; they will derive their own internal satisfaction. Sadler (1995: p. 124) in concurring with this point explains that learning in organisations can occur in two ways:

(1) Direct (immediate)
(2) Indirect (deferred)

Direct learning is the type that has traditionally been used, and consists of training and lectures by in-house or external experts to assist employees with carrying out particular aspects of their jobs.

Indirect learning involves the transfer of knowledge through the use of stored information contained in an organisation's systems (including the quality management system), policy statements, and most notably, the culture that members exhibit. In essence, the challenge to management wishing to instil the ethos of organisational excellence, is to create an environment in which all employees are encouraged to seek ways of engaging in indirect learning, rather than waiting to be told what they should learn.

Sadler provides the following diagram (see Fig. 4.6) that indicates the various processes and influences which are involved in a learning organisation.

Peter Senge, one of the most well-known proponents of organisational learning, contends that there are five 'disciplines' that support the methods being used. These are:

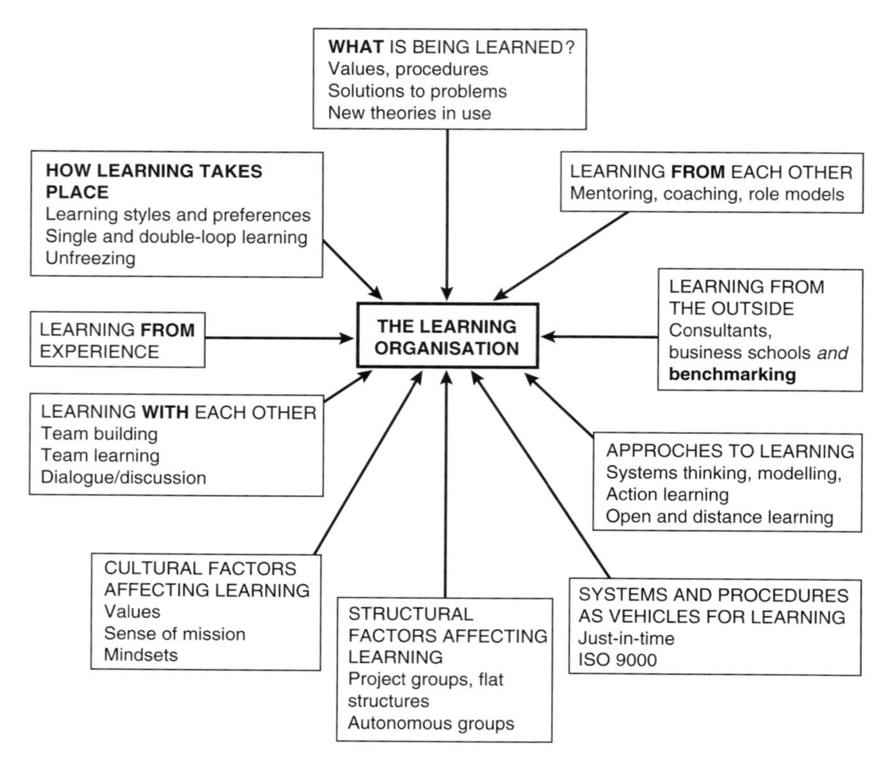

Fig. 4.6 The processes and influences that 'create' learning organisations.

(1) *Systems thinking* which according to Senge *et al.*, are the 'forces and relationships that shape the behaviour of systems' (Senge *et al.*, 1994: p. 6). As Deming advised, it is the duty of management to understand how an organisation's system operates and the limits to its growth and development.
(2) *Personal mastery* whereby individuals are encouraged to have enquiring minds so as to continually seek better ways of doing things. As Crainer describes, people in a learning organisation are expected to 'live life from a creative rather than a reactive [perspective]' (Crainer, 1996: p. 238).
(3) *Mental models* are the way that commonly held assumptions are challenged. As such, people are encouraged to develop new models which allows them to consider radical ways of carrying out day-to-day activities.
(4) *Shared vision* is, according to Senge, vital. As Sadler asserts, in any organisation where there is a shared vision among its

members, 'people are capable of outstanding achievements' (Sadler, 1995: p. 128).

(5) *Team learning:* this of course follows on from what has been described previously in this chapter – that teamwork is a vital component of improvement. Indeed, as Senge *et al.* contend, learning by groups will always be greater than the 'sum of individual members' talents'(Senge *et al.*, 1994).

Argyris (1993), another well-known proponent of learning, is similar to Sadler, in that he believes that learning takes place at two levels:

(1) Single-loop learning
(2) Double-loop learning

As he explains, in an organisation that provides customers with a product or service that is predictable, it is appropriate to provide employees with instructions that are mandatory and must be learned by rote. An example of this is the McDonalds food chain. Employees are expected to provide food which, despite some variation (for particular regional reasons), tastes consistent throughout the world. However, as Argyris explains, double-loop learning occurs where employees are allowed to question the accepted method of doing things. The emphasis, therefore, is to encourage employees to continuously seek alternative methods, regardless of how long they have existed.

In addition, Argyris draws attention to what he calls managers' espoused theories and their theories-in-use. The former, he explains, is what managers describe *should* go on in the organisation. The latter he explains, is what people who carry out day-to-day operations describe as *actually* happening. According to Argyris, any deviance between espoused theories and theories-in-use frequently results from managers being out of touch with what goes on. Therefore, he suggests, managers should be prepared to realign their espoused theories to match reality. If improvement is subsequently required, then at least everyone is aware of what *really* happens – not an artificial view of what managers *believe* may happen. This reinforces a point that Sadler makes – the need for honesty in achieving an organisation where the desire to engage in learning becomes really embedded.

In order to assist those who attempt to create change management, Adams, Hayes and Hopson (1976) propose a model that has seven phases (see Fig. 4.7). The seven phases and typical characteristics are as follows:

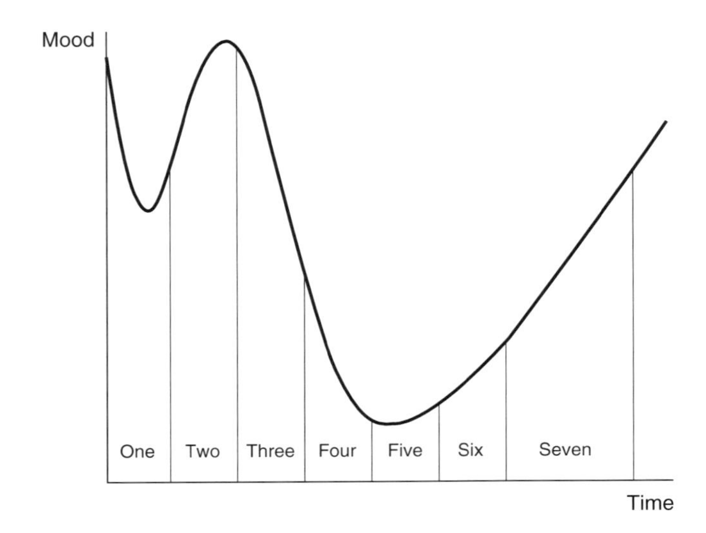

Fig. 4.7 The Adams, Hayes and Hopson feedback acceptance model.

(1) *Phase one: shock*. During this phase, people are, literally, shocked by the fact that things are starting to change. It is common for people to engage in activities that add nothing; they are 'time wasting'.
(2) *Phase two: denial*. During this phase, people have moved on from shock and are resigned to the change. However, rather than embracing it, they either prefer to pretend that it won't affect them, or become resistant to the change process.
(3) *Phase three: disillusionment*. During this phase, people are likely to have become disaffected by the change process. There is a real danger that if they are allowed to continue like this for any length of time, the process will be terminally damaged.
(4) *Phase four: letting go*. During this phase, people will have begun to accept the change process and may, possibly, begin to see benefits to them by engaging with it.
(5) *Phase five: testing*. During this phase, people will start trying things out to see what effects may occur. Even though it may be unsystematic, it will show that there is definite support for what is happening.
(6) *Phase six: consolidation*. During this phase, people will be much more willing to co-operate and exchange ideas so as to ensure that their efforts are carried out in a co-ordinated and systematic way.

(7) *Phase seven: internalisation.* By this stage, the change process will have become 'normal'; people will accept that whatever the end objectives were, they are now viewed as 'the way that things happen around here'.

As will be described in subsequent chapters, most especially Chapter 7 which describes the use of the EFQM Excellence Model, it is important to recognise these phases as being a normal part of the 'emotional turmoil' that organisational change involves. Thus, the first three phases – shock, denial and disillusionment – may threaten some people and cause them to overreact. As those who have examined what change management actually involves have found, these three phases are necessary to allow people to come to terms with the need to 'let go' of the traditional way of doing things (phase four). By so doing, it is then possible to encourage people to attempt to try different approaches to carrying out routine tasks (testing – phase five), and if success is secured, to consolidate (phase six), and as a consequence, ensure that continuous improvement becomes part of the organisational culture (internalisation – phase seven).

Summary

This chapter has stressed the importance of recognising the issue of organisational culture. In particular, those wishing to use benchmarking to produce organisational excellence must understand the following:

- What organisational culture is
- What influences the development of organisational culture
- The importance of senior and middle managers in changing the culture of an organisation
- That people who carry out day-to-day activities must be encouraged (motivated) to engage in changing their attitudes and behaviour to produce improvement
- The impact of effective teamworking
- That during any process designed to change the culture of an organisation, there will be upheaval and emotional turmoil (this needs to be managed)

CHAPTER FIVE

THE USE OF CRITICAL SUCCESS FACTORS, PROCESSES AND SYSTEMS IN BENCHMARKING

Objectives

Previous chapters have explained the theory that underpins the need to use benchmarking as a tool for achieving organisational excellence. However, unless there is a starting point it will be difficult to know how much progress has been achieved. This chapter describes how to put in place these starting points:

- Critical success factors and key performance indicators
- How these are used to measure the rate of improvement

In addition, this chapter will explain why it is essential to understand what processes are, and how they can be mapped to assist in identifying areas for improvement. It will also describe how quality systems such as ISO 9000 (BSI, 1994) can be invaluable in keeping processes 'under control' (something Deming stressed was essential, and which is an aspect that organisations must demonstrate as part of a submission for assessment of the EFQM Excellence Award - see Chapter 7).

5.1 *Where to start from*

In order that an organisation can attempt to use the technique of benchmarking, it is assumed that this decision is one that will have emanated from the senior managers. Even though much is made of the need for people to be more involved in the day-to-day decision-making of how tasks are carried out, and therefore to be more involved in the effort to suggest ways of improving, senior managers must be prepared to tell people what is expected.

In communicating to people what is expected of them, it is important that senior managers are capable of providing a vision of where the organisation is going and therefore what its corporate aims are. In providing this vision, they are expected to be able to articulate what is commonly known as the *mission*. This mission will describe what senior managers believe needs to be achieved to attain particular objectives. In particular, they will need to understand what is currently done, and how, using benchmarking, the ability of this organisation can be enhanced to do this. In order to start this process it is essential to define the *critical success factors*, and how, subsequently, these will be used to identify the key processes which need to be improved. It is these key processes which will form the 'building blocks' upon which the process of benchmarking will be built so that the critical success factors are attained.

5.2 *Critical success factors and key performance indicators*

5.2.1 Critical success factors

Critical success factors (CSFs) and key performance indicators (KPIs) were described in Chapter 1 and as the next chapter will explain, one of the main determinants of success in business is the ability to know how satisfied your customers are with what you provide. There are various ways of eliciting data on customer satisfaction. Ensuring that those who consume/buy[29] choose your product or service is, of course, the ultimate goal of any organisation. If an organisation does not satisfy customers, for whatever reason, they are likely to take their custom elsewhere.[30] This objective represents what can be regarded as the most important CSF. As Bendell *et al.* explain, CSFs:

> ... represent a small number of key indicators [by] which, provided they are showing satisfactory progress towards targets, the organization will generally be seen to be succeeding in its task of quality improvement. (Bendell *et al.*, 1997: p. 87)

CSFs, therefore, are vital for managers engaging in improvement of their organisation, as they will indicate how much progress is being made in particular areas. As such, the choice of CSFs is important. Their choice also poses a danger especially if those who choose them are unfamiliar with the capabilities and resources that the organisation possesses. The consequence of choosing CSFs and setting targets that are impossible to achieve may prove demoralising and potentially destructive to further improvement activities. For this reason, there is an argument that CSFs must be chosen subsequent to the identification of key processes. However, at the outset of the benchmarking process, using general CSFs which indicate *what* the organisation hopes to do should not cause problems. The procedures that are used to control the key processes (see below) represent *how* it will be possible to realise the attainment of the CSFs. Oakland makes the important point that a CSF should represent a 'balance of strategic and tactical issues' (Oakland, 1999: p. 27). As he stresses, the tactical issues are concerned with the day-to-day processes that are carried out.

So for instance, CSFs that might be chosen at the initial stages of the benchmarking process might be:

- This organisation will improve the satisfaction levels of its customers
- This organisation will ensure that the competency and skills of its workforce are raised
- This organisation will attempt to become recognised as one of the leaders in the use of best practice
- This organisation will develop the relationship it enjoys with key suppliers and subcontractors

Having created these statements, they should achieve the following:

(1) Be positive in intent (it is important to avoid vague wishes which can be misinterpreted)
(2) Represent a balance between what is desired and what can be achieved (not overambitious or too easily achieved)
(3) Make it possible for managers and employees to understand which departments and processes will need to be targeted in order to attain success in particular CSFs
(4) Make it possible to quantitatively measure progress in their achievement – this requires the use of what are called key performance indicators

5.2.2 Key performance indicators

Oakland makes the point that whilst CSFs are important in terms of setting direction for everyone in the organisation, they can, without specific targets to be achieved, be seen to be 'loose statements'

(Oakland, 1999: p. 27). In effect, these targets – what are commonly known as key performance indicators (KPIs) – represent the measures of progress in achievement of the CSFs. So, like the long-distance runner who trains daily for a marathon by measuring his/her rate of progress, an organisation should be able to collect data concerning each CSF. Doing this will allow constant monitoring to be carried out and therefore, as a result, analysis of the effects of any changes introduced to be considered. It is at this point that the managers responsible for having selected the CSFs – the *whats* – can actually achieve their goals. This, in effect, means that the CSFs must be operationalised into *hows;* a task that requires the identification of processes that relate to each CSF. Therefore, it will be necessary to identify those people who are most closely associated with these processes.[31] The main objective at this point will be for someone from senior management who was responsible for selecting the CSFs to facilitate an exploration of these processes. As a result, it will be possible to:

(1) Create an effective team to consider how this activity can be carried out
(2) Agree the actual measures that can be used to monitor the KPIs
(3) Provide a method which enables the consistent collection of appropriate data
(4) Suggest possible changes that will create opportunities for improvement
(5) Review progress and, where necessary, alter the targets

The next stage in this activity is probably the most crucial in determining the success of benchmarking for best practice: the need to understand processes.

5.3 *The importance of understanding processes*

The word process is one that occurs frequently when describing quality management or benchmarking. As Brown suggests, the word process seems to cause problems to people involved in non-engineering organisations (Brown, 1993: p. 123). The word *process* is defined by Macdonald in the *Chambers Twentieth Century Dictionary* as: 'a series of actions or events: a sequence of operations or changes undergone' (Macdonald, 1972: p. 1070).

Oakland defines a process as being the 'transformation of a set of inputs, which can include actions, methods and operations, into desired outputs which satisfy the customer needs and expectations' (Oakland, 1999: p. 55). As those who advocate improvement using QA, TQM or benchmarking advise, any person or organisation will, in the course of providing products or services (regardless of context), carry out any number of primary and sub-processes.[32] Thus, they argue, if an organisation wishes to raise the level of satisfaction that a customer derives from the consumption of a product or service, it is essential that the processes that are involved are fully understood. Sylvia Codling uses the analogy of an onion (Codling, 1992: p. 56). As she describes, an onion appears to be a complete entity. However, as anyone who has peeled an onion will quickly discover, it really consists of many layers. This, she asserts, is like any major process that exists in an organisation.

Figure 5.1 shows the main inputs and outputs which, according to Oakland, are part of a process (Oakland, 1999: p. 56). As this diagram indicates, in order to ensure consistency of output, it is necessary to control the inputs. This is exactly what Dr Deming argued with regard to managing quality – that it is essential to control the variability of processes. However, despite the apparent simplicity that this concept suggests, it is surprising how little people actually know about what any process involves, or why they are carried out in a particular way. In considering processes, it will be necessary to break processes down into a manageable size. This normally involves considering sub-processes.

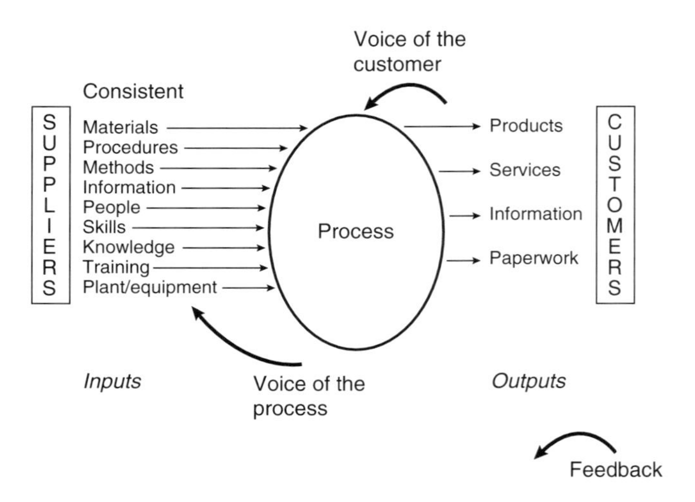

Fig. 5.1 A typical process

Additionally, the sub-processes can then be divided into activities and tasks. Activities are normally considered to be carried out by groups (teams) whereas tasks are assumed to be achieved by an individual. As has been stressed already - particularly where the philosophy of Dr Deming was described - in improving processes (and their constituent parts), it is crucial to involve those people directly involved. Unless such people are involved, Deming and others would argue, how is it possible to ensure the following:

- That information concerning the processes is accurate
- That comparison with processes of organisations which are 'excellent' are understood by those who will be responsible for implementing any changes - in addition, any changes made to the processes will be carried out with the agreement of those who will be directly affected
- That potential improvements suggested by those who carry out tasks/activities are incorporated

Thus, if being involved in benchmarking achieves nothing else, the need to document processes (and procedures[33]) for carrying out operations, will - if done accurately - clarify what actually goes on. For reasons that only become apparent upon detailed examination, what is frequently suggested to be 'standard practice' bears little in relation to what actually goes on. Until any such differences between what is described and what goes on with respect to processes are reconciled, it will be difficult to institute changes that will result in improvement. As may become apparent, there may be processes that are being carried out that owe more to tradition than an emphasis on value-adding.[34]

5.4 Process mapping: 'the metaphor of the cup of tea'

This is a technique that involves describing - using pictorial representations - how a process is carried out. As should become apparent on examination of the process, there will normally be a number of steps. Each step may involve the completion of a specific task that is essential before subsequent tasks can commence; there is sequential dependence. Some tasks may be capable of being carried out simultaneously. Anyone who has ever studied a critical path chart will see this sort of sequence very clearly.

In order to illustrate how a simple process can be mapped, the reader should consider what would be involved in making one cup of hot tea. In order to achieve a result that would be considered 'satisfactory', the following steps or conditions must be achieved.

(1) That it is possible to find the basic materials required – teabag, water, milk and, if necessary to provide the right taste, sugar
(2) That there is essential equipment (cup, kettle and spoon) and the means by which to heat the water
(3) That the water is placed into the kettle
(4) That the heat is turned on
(5) That there is a method of knowing when the water is boiled (either manually or the use of a whistling kettle)
(6) That the water is successfully poured into the cup containing the bag
(7) That after a reasonable time (depending on the desired strength of tea) the bag is removed using the spoon
(8) That the spoon is used to stir a quantity of milk and sugar that the consumer has decided is necessary for an acceptable taste

As you might reasonably say, this an overly technical way of considering what is a simple process that we frequently do without thinking. That is the point about understanding what goes on in carrying out day-to-day activities in an organisation: because they are done so often, no one bothers to think any more. As a consequence, the attitude will probably be, 'Why should we do it any differently? It's always been done like that'.

So, in considering the cup of tea, how should it be mapped? Figure 5.2 would be a reasonable representation of the eight steps being carried out in a linear sequence. As previously suggested, this is such a mundane process that no one usually considers it necessary to think about what goes on. Indeed, it is usual to do other things while the kettle is boiling, such as watching television, or making a sandwich. However, what is important about describing this apparently simple process is that by breaking it into discrete steps, it is possible to think about changes in one or more of them that may create certain improvements. The nature of such improvements may be, for example, in the time it takes to complete the process of achieving the cup of hot tea. By examining the map in Fig. 5.2, it will be clear that the step which takes longest is that in which the water is boiled. Therefore, if the whole process is to be completed more quickly, it will be necessary to change the method of boiling the water by using a microwave or 'high-speed' electric kettle.

In this simple example there is only one person, and it is that

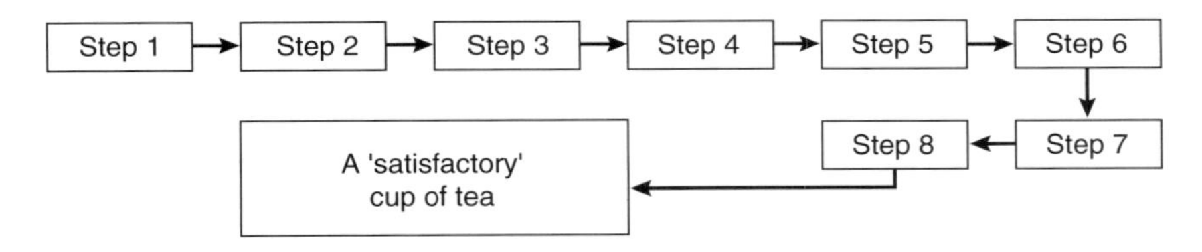

Fig. 5.2 A process map for making a cup of tea.

person who normally decides how long the teabag should remain in the cup. This, as anyone who has made a cup of tea this way will have discovered, is dependent upon a certain amount of judgement (time and the colour of the water). However, if one wanted to ensure that the taste was consistently strong or weak – depending on the desired taste – it would be necessary to consider specifying the following:

(1) The strength of the bags being used (brands differ)
(2) The length of time it takes to create a particular taste
(3) The amount of milk and, if required, sugar that must be added

Therefore, if someone else was required to achieve the satisfactory cup of tea, it would be necessary to provide exact procedures that this person could understand, and be capable of complying with. This is the objective of creating a documented quality system. What this involves, is described in the next section.

Using the example of the cup of tea as a process demonstrates that such an apparently trivial task is more complex than it might appear, and that the end result is dependent upon variables that occur in certain steps, and which, in turn, influence subsequent steps. This is a crucial point, because as Codling points out with respect to processes, it is the 'boundaries' between each step that are frequently the 'root of problems or dysfunctions' (Codling, 1992: p. 61). In any organisation where processes far more complicated than making a cup of tea are required to create the product or service, how can it be possible to see where improvement is possible without first mapping the process? Clearly, the more accurate the information that is elicited to provide the process map, the easier it will be to identify potential areas for improvement in the overall process. It is important to stress that the key to success in doing this, is to consistently do the following things:

- Know what the 'end' customer's requirements are, and what is needed to achieve them

- Know the requirements of those involved in each next step of the process, so that those in preceding steps can achieve them using adequate resources
- Use benchmarking in order to learn how best practice can be implemented in the process
- Using procedures which are easily understood by users, document how particular tasks/activities should be carried out

As the next section explains, documented procedures are what normally constitute the basis of a management system, particularly a quality system such as one that conforms to ISO 9000 (BSI, 1994).

5.5 Quality systems and procedures

In BS EN ISO 8402 a quality system is defined as '[the] organisational structure, procedures, processes and resources for implementing quality management' (BSI, 1995: p. 26). As the cup of tea example demonstrates, subsequent to having made decisions about the way a process is carried out (*organised*), and the materials and equipment that are required (*resources*), it is still essential to ensure that the person(s) given the task of achieving the desired objective are capable of carrying out particular tasks in a way that is both competent and consistent. The former element - competency - is assumed to be taken into consideration when employees are selected.[35] Additionally, the use of training is considered to be important to ensure that employees are aware of how to use new technology and skills. However, despite the view that people can be allowed to use their skill to decide how a particular task should be performed, it is normal to encourage the use of procedures that describe 'standard' practice.[36]

Procedures, therefore, are considered to represent guidance to people on the methods/tools they should use for tasks or activities that are normally considered to be routine. Thus, provided that these procedures have been created on the basis of 'capturing' what is *agreed*[37] to represent best practice, compliance with them will be considered to be unproblematic. There are those who argue that a documented quality system, precisely because it involves using written procedures, is unlikely to allow an organisation to achieve improvement (Seddon, 1997). There is a certain amount of validity to this argument. From previous research (McCabe, 1999), it can be seen that this has brought about the use of what are perceived to be bureaucratic procedures. This is due to the following reasons:

- Procedures that are created without the agreement of those who will be expected to comply with them
- Procedures that are 'over-written' because of too many words, or have technical, jargonistic or misleading language
- The use of the quality system turns into what has been referred to as a 'paper tiger' (operating the system becomes a bureaucratic exercise in itself)
- Because the operation of the quality system is audited, there is a tendency for people to 'merely do enough' to comply with the procedures – doing any more than this is regarded as being a waste of effort

This state of affairs has been seen to represent an 'audit culture'. Accordingly, the use of a quality system becomes something that disables people rather than enabling them. This does not have to be the case. Any procedure must explicitly tell the user what they should do (and if for any reason this involves doing something that is not obvious, why). The aim is usually to ensure that by following the procedure the result(s) of the process are consistent. It is for this reason that the use of diagrams or pictures is an entirely sensible way of providing procedures.[38]

Clear and informative procedures are also found to be vital if the person carrying out this process is taken ill or takes a holiday (something that frequently happens). In this situation, it is possible to instruct another person to carry out the task with little risk of the end result being any less consistent than before. As excellent organisations have discovered, ensuring consistency in the standard of the product or service that customers receive is essential; any failure to meet expectations is unlikely to attract sympathy, regardless of the cause.

The problem of procedures that require too much paperwork can usually be overcome by the use of simple methods of checklists. For instance, often when going into the toilet of a restaurant it may be noticed that there is a sheet of paper that records (using a tick) the number of times that it has been cleaned. Using a tick or initials of the person carrying out the task is an efficient method of providing a record that allows for immediacy in auditing. Clearly, what is important as far as the customer is concerned is that the task performed matches their level of satisfaction. On one memorable occasion in the toilet of a restaurant, the control could see that they had supposedly been cleaned only an hour before but still smelled awful. Even though there may have been a reason for this, it is strongly suspected that this was a case of 'tick box syndrome' – that is, that the cleaner(s) were dishonestly pretending to have done the task. If this is allowed to happen, the quality system will soon fall into disrepute.

5.5.1 The use of quality systems in continuous improvement

Oakland explains that the procedures of any quality system form what he suggests are like the slats on a rotating drum, the objective of which is to create a 'delighted customer' (Oakland, 1999: p. 86). As he describes, the managers must ensure that all of the necessary elements are in place to continuously turn the drum. Crucially, he stresses, it is important that the procedures are carried out with a desire to monitor (measure) their effectiveness in attaining the objectives of the process. Thus, he asserts, any changes that are implemented to create improvement in the overall process must be incorporated into the documented procedure immediately:

> The overriding requirement is that the systems must reflect the established practices of the organisation [and] improved where necessary to bring them into line with current and future requirements. (Oakland, 1999: p. 88)

The issues of reflecting what really goes on was something that was described in Chapter 4, which explained what organisational culture involves. In particular, there is the problem of 'espoused theories' (what is suggested happens) being different to 'theories in practice' (what those who carry out activities consider really goes on). If those who must use the quality system do not believe it reflects what really goes on, it will lack credibility. As a consequence, the likelihood of audits indicating that procedures are being ignored will be increased; something which third-party assessors must draw attention to in their reports. Therefore, to ensure that users do not perceive the quality system to be lacking in credibility, what it contains must both be a reflection of 'the way that things are done here' (a simple definition of culture), and provide information that is believed to be useful.

The effect of using benchmarking will be to encourage and assist employees to think of different – even radical – ways of carrying out their day-to-day duties. In effect, there is the desire to support a culture in which people view the use of procedures as being a process in itself, and most especially, one in which the aim is constant improvement. This approach is consistent with the philosophy

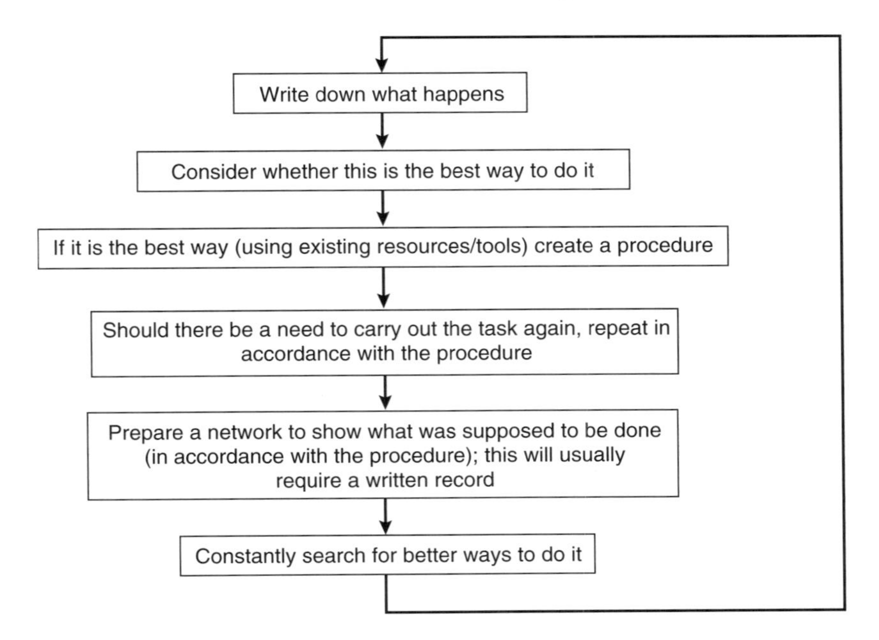

Fig. 5.3 The search for continuous improvement in procedures.

of Deming who recommended that every task should be subject to an improvement cycle (see Fig. 5.3).

As this diagram indicates, the cycle is continuous, and therefore the result of diagnosis of action that has already occurred should result in alteration in how future action is planned. Deming argued this approach should be adopted by every person. As a result, he suggested, improvement in the overall process must inevitably result. Figure 5.4 adapts Deming's PDCA cycle to ensure that the desire to improve everything drives the whole system, including, it should be stressed, those who are external to the main supply organisation.

As this diagram also shows, this model draws attention to subcontractors and suppliers. This is an important point to remember, because as any customer will discover if they examine the process by which the product or service they have purchased/consumed occurred, it is extremely rare that only one organisation was involved. Therefore, the quality system must actively engage in improvement, not only as a result of what the external customer tells you, but also by ensuring that suggestions by those who create components or services are included. The desire to create closer relationships with key suppliers and subcontractors is commonly

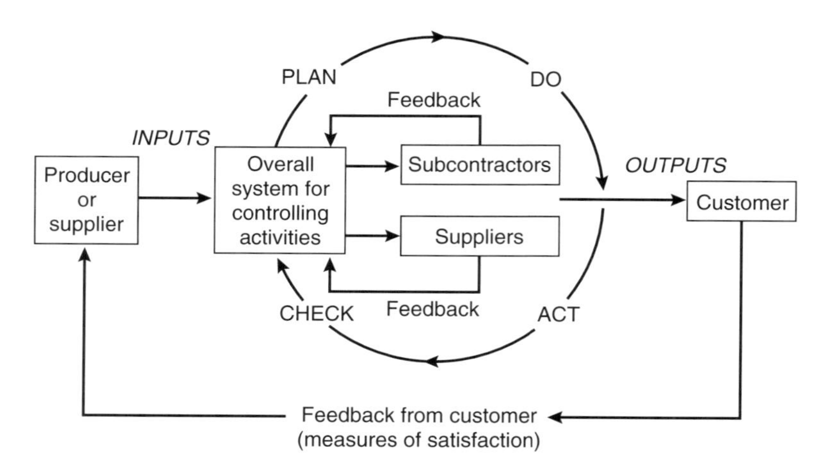

Fig. 5.4 Overall improvement in total processes.

known as *partnering*, something that can equally apply to the relationship that an organisation has with its *key* customers. The benefit of this sort of relationship is that, like getting to know close friends, it becomes much easier to understand needs, expectations, ambitions and most especially, potential weaknesses.

All parties have a vested interest in improving all aspects of the processes which contribute to the final outcome - customer satisfaction. Returning to Deming's cycle which is in Chapter 3, he believed that the result of better quality is to make the customer want to come back for more. This will benefit all. Therefore, if those who supply you can assist in identifying ways of improving existing methods of working, the likely consequence is to improve the satisfaction levels of the customer. Clearly, if better methods are found to be possible, they should be incorporated into the overall system. The objective, it should be noted, is for every person to constantly search for ways to enhance the satisfaction levels of the customer. As Chapter 7 will explain, customer satisfaction is the part of the EFQM Excellence Model which has the potential for the highest score. Therefore, the ability to benchmark and measure customer satisfaction is very important. It is this issue that the next chapter considers.

Summary

In order to be successful in benchmarking, it is essential that the senior managers of an organisation decide upon what its mission is. This is crucial to provide all employees with a clear idea of the objectives that are to be attained. It should therefore be communicated to every person in the organisation in language that is clear and unambiguous.

- Subsequent to the mission having been decided upon, it is necessary to translate it into what are known as critical success factors (CSFs). CSFs provide a focus for what people will be aiming to achieve in order to ensure that the mission is successful within the period that has been allotted.
- Corresponding with CSFs are what are known as key performance indicators (KPIs). KPIs are the measures that are used to provide targets against which progress towards achievement of CSFs can be assessed. As such, the development of KPIs should be achieved in consultation with those people who are directly involved in the carrying out of processes (see below)
- In order to ensure that the improvement effort can be operationalised, it is essential that each CSF and the KPIs associated with it are considered with direct respect to the day-to-day processes that are carried out.
- Unless what happens within the processes can be changed and made to occur more effectively, the likelihood of being able to show that the KPIs associated with the CSFs have improved will be reduced. As a result, it is necessary to consult those people who have direct involvement with the sub-processes, activities and tasks.
- In order to assist the exploration of the process, sub-process, activities and tasks, it is recommended that a technique called process mapping should be used. This technique requires that rather than describing what goes on, a pictorial representation is used. As a consequence, it should be simpler to identify boundaries and potential areas that can be improved.
- Any changes that are made to the process and which result in improvement are carried out consistently by the alteration of procedures which directly concern the execution of activities and/or tasks.

CHAPTER SIX
BENCHMARKING CUSTOMER SATISFACTION

Objectives

The concept that customers' satisfaction should be seen as something new is, in reality, mistaken. Any study of history will show that successful commercial relationships were built upon the ability of parties engaged in transactions to derive satisfaction. One quote from the person who founded the department store Selfridges (Harold, 1857–1947) summarises what this chapter seeks to explain:

'If you mean to profit, learn to please'.

Dr Deming, someone who is credited with assisting the post World War II Japanese reconstruction, believed that quality is not just simply about *satisfaction* but about *delighting* customers. Deming's belief has been echoed in the pronouncements of more contemporary writers on quality and customer service. For instance, Schonberger in his book *Building a Chain of Customers*, argues that organisations which have achieved what is called 'world-class' status (see Chapter 7) have a common characteristic – a dedication to customers' satisfaction (Schonberger, 1990: p. 1). Scherkenbach (1986) warns that unless customers are made the number one objective in business, the very survival of the organisation becomes endangered.

The implication of this statement is that every organisation should do everything it can to ensure that it knows exactly what its customers want, and more especially, how good the products or services they consume are. As a direct consequence, it should be possible for managers to consider potential changes that can be made to processes and methods used in production or service delivery. This will ensure that satisfaction levels are, at the very least, in the short to medium term, maintained, and in the medium to long term, improved, so that the organisation is perceived as 'giving the best that money can buy'. In order to do this effectively it is essential that some form of measurement of the level of satisfaction is carried out. This chapter examines the following aspects of benchmarking customer satisfaction:

- The methods that exist in order to measure customer satisfaction
- Why an organisation should put the measurement of customer satisfaction at the heart of its corporate mission
- How, by dedicating effort to ensuring customer delight, organisations can both increase market share and reduce 'wasted' costs on advertising
- The importance of a technique called *relationship marketing* in maintaining customer loyalty

6.1 The paradigm shift in customer value strategy

In their book *Beyond Total Quality Management – towards the emerging paradigm*, Bounds *et al.* make a comparison between what they call the 'old' and 'new (emerging)' paradigms of customer value strategy (Bounds *et al.*, 1994: p. 29). The new paradigm, they explain, requires organisations to engage in change in the following areas:

(1) *Product or design.* To shift from a state where the focus is 'internal', and that it is only possible to 'sell' what the organisation has the current capabilities to produce to one where the focus is external, and as a result, '[produce] what customers want' (Bounds *et al.*, 1994).
(2) *Quality.* To move from the belief it is simply a matter of 'meeting specification' and making 'trade-offs' between quality, cost and schedule to a desire to 'seek synergies' with respect to these components (Bounds *et al.*, 1994).
(3) *Measurement.* To change from the use of internal measures which are 'not necessarily linked to customers' to the implementation of all measures being 'linked to customer value' (Bounds *et al.*, 1994).

The following three aspects of customer value strategy summarise what benchmarking customer satisfaction is about:

(1) Find out what they really want
(2) Make sure, at the very least, that what you give them performs to their expectations (the aim, of course, is to try and surpass their expectations)
(3) Continuously measure what you have done in order to ensure that every aspect of what they receive is giving 'value'

The challenge that managers in organisations face, therefore, is in being able to effectively 'know' what customers really want, and even more importantly, to exceed what they will find acceptable. As the next section explains, benchmarking customer satisfaction is no longer something that any organisations should *merely consider*, but instead, must *carry out* as a matter of routine.

6.2 The benefits of retaining customers

Anyone who has ever tried selling anything will know that unless people are queuing to purchase what you sell, a great deal of effort is frequently required to convince them. This is true regardless of the size of the organisation or the context in which it operates.[39] As organisations increasingly find, more effort and cost is required to attract new customers. However, various studies have found that if organisations concentrate their efforts on existing customers, the dividends can be considerable. For example, Zairi refers to a survey conducted by Bain and Co. which concludes that the capability of profit-making organisations to retain customers is one of the biggest influences on their 'bottom line':

> The study concluded that the ability to reduce customer defections by 5 per cent can lead to increases in profit levels by anything from 25 per cent to 85 per cent. (Zairi, 1996: p. 59)

Research that Glen Peters refers to, similar to that which Zairi cites, suggests that retaining customers is always beneficial (Peters, 1994). In particular, he believes that in an increasingly competitive marketplace, the costs of replacing lost customers will significantly impact on the overheads. (His research shows that every 2% increase in customer retention will lead to a 10% reduction in overheads.)

The implication is therefore clear: ensuring that the customers you currently have remain loyal makes good business sense. Thus, in an industry like construction where profit margins tend to be so low (more than 5% being seen to be exceptional), customer retention represents an effective method of increasing this aspect of the business. The problem, therefore, is in knowing what to do, and more especially, how.

6.3 Factors that must be considered when measuring customer satisfaction

Zairi, in explaining how to benchmark customer satisfaction levels, draws attention to the following general principles that must be applied (Zairi, 1996: p. 189):

- The metrics[40] that are used to monitor the customer's satisfaction are accurate
- These metrics should be sensitive (they clearly indicate a causal relationship)
- The results should be capable of comparison with direct competitors
- The process must be regular and continuous
- Whatever method is used, it should be simple, and the results easily communicated to those who are affected

These principles, Zairi argues, provide the basis upon which all improvement efforts should take place. As a consequence, therefore, the key to success in pleasing customers is the ability of the organisation – i.e. its employees – to be able to translate changing customer expectations into the actual product or service in minimum time. If an organisation is able to achieve 'customer-inspired' changes to its products or services faster than those who are in opposition, the effect will be to enjoy, in the short term at least, *competitive advantage*. This principle is one that Toyota have refined in applying the concept of lean production. As has been explained previously, an organisation must ensure that those who work on its behalf, and as a result either create the product or service or interface with the customer, are capable and motivated to strive for excellence. Zairi believes that there are a number of similar characteristics found in organisations that are customer-focused and which, he argues, allow them to become 'excellent':

- That employees are aware of the results of benchmarking customer satisfaction

- That employees fully understand the consequences of trying to ensure the customer derives delight
- That employees are encouraged to use their initiative, and where necessary in the interests of customers, to deviate from 'standard practice'[41]
- A culture of seeking ways to - as Peters and Waterman advised (1982: pp. 156–200) - 'get close to the customer'
- Creation of teams that co-ordinate activities which result in the totality of the process being absolutely dedicated to customer satisfaction

What this may involve has been described in previous chapters. However, what is important is that all employees - most especially those who come directly into contact with customers - should be trained to be constantly aware of any information that can be gleaned about satisfaction (and, of course, non-satisfaction) levels. Whilst the next section deals with formal models that can be used to elicit data on customer satisfaction, the ability of people 'at the sharp end' who are able to act as 'eyes and ears' will be vital in becoming aware of the possibility of customers considering taking their trade elsewhere.[42] As studies alluded to earlier stress, intervention at this point that results in maintaining their custom can produce potentially great financial dividends.

6.4 *A selection of models that can be used to carry out benchmarking of customer satisfaction*

This section describes some of the standard models that exist in order to carry out benchmarking of customer satisfaction.

6.4.1 The SERVQUAL model

One of the most widely recognised models for benchmarking customer satisfaction is known as SERVQUAL. This model was developed by Parasuraman *et al.* (1988). The SERVQUAL model attempts to measure the differences (*gaps*) that exist between what a customer expects and their perception of the level of quality they actually receive. As Parasuraman *et al.* believe, 'the criteria used by consumers is assessing service quality to fit ten potentially overlapping dimensions' (Parasuraman *et al.*, 1988: p. 17). These are shown in Fig. 6.1.

1 *Tangibles*
- ☐ The physical attributes that exist in a product or service

2 *Reliability*
- ☐ Whether the product or service performed in accordance with what would have been expected

3 *Responsiveness*
- ☐ The willingness of the service provider to respond to requests which may require particular alterations

4 *Communication*
- ☐ The ability to talk to customers in a way that they can understand

5 *Credibility*
- ☐ The honesty and esteem that the provider extends to customers

6 *Security*
- ☐ This can be physical, financial or concern confidentiality

7 *Competence*
- ☐ The ability of the employees of the provider to deliver the service

8 *Courtesy*
- ☐ The politeness and respect which staff in the provider organisation give to their customers

9 *Understanding/knowing the customer*
- ☐ The ability of the provider to appreciate what their customers really want

10 *Access*
- ☐ The ease with which customers can communicate with staff in the provider organisation

Fig. 6.1 The ten SERVQUAL criteria used by consumers to assess service quality.

Following research work by Parasuraman *et al.*, these ten dimensions were refined and reduced to five that appear in the SERVQUAL model. As these five (which are listed below) show, the first three (tangibles, reliability and responsiveness), are unchanged from the original list, whereas the last two are a combination of the other seven original dimensions.

The five dimensions of the SERVQUAL model

The definitions of the five dimensions provided by Parasuraman *et al.* (1988: p. 23) are as follows:

(1) *Tangibles.* Physical facilities, equipment, and appearance of personnel
(2) *Reliability.* Ability to perform the promised service dependably and accurately
(3) *Responsiveness.* Willingness to help customers and provide prompt service
(4) *Assurance.* Knowledge and courtesy of employees and their ability to inspire trust and confidence
(5) *Empathy.* Caring, individualised attention the firm provides its customers

Implications of using this model

As Parasuraman *et al.* point out, these five dimensions provide what they admit to be a 'basic skeleton' (Parasuraman *et al.*, 1988: p. 30). Therefore, they advise, it will need to be 'adapted or supplemented to fit the characteristics or specific research needs of a particular organisation' (Parasuraman *et al.*, 1988: p. 31). As they also advise, the SERVQUAL model works best when it is used together with 'other forms of service quality measurement'. The most important thing about SERVQUAL, the authors stress, is that it is used systematically over a long period of time, and that any gaps that are identified are dealt with immediately. This is the most important thing to emerge from their study: that there is a desire to find measurable gaps between customer expectation and their perceptions of the levels of service they receive. By constantly attempting to reduce such gaps, it is highly probable that the excellence this organisation provides will be improved.

6.4.2 The Christopher and Yallop model

This model is named after the people who developed it. Like Parasuraman *et al.*, Christopher and Yallop (1991) argue that in the contemporary market the most important influence on a customer is the perception of the service they receive. Enhancing customer service, they believe, coupled with a differentiation of the core product or service that is provided, allows an organisation to enjoy competitive advantage. As Fig. 6.2 shows, Christopher and Yallop show service characteristics to be what they call 'product [or alternatively service] surround'.

What this diagram clearly shows is that whilst the basic product or service should be as good as possible, because the surround can

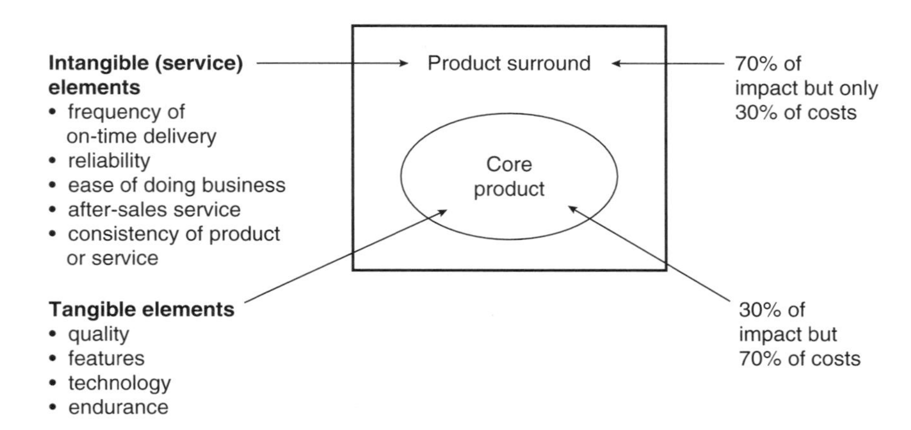

Fig. 6.2 The addition of service elements to enhance core product.

create 70% of the impact, it is by adding to this that market differentiation can be established. However, as proponents of this model stress, what constitutes the product surround are intangible; they are highly perceptual, and therefore, subjective. The consequence of understanding this, they argue, is that the method used to elicit data that concerns customer service must be sensitive to appreciating that these perceptions are what the customer feels is important to them. In order to attempt to measure the perceptions that customers have of the service they receive, Christopher and Yallop suggest a four-stage process model:

(1) Identify the elements of customer service which are perceived as being crucial by customers
(2) Find a way to understand the importance that is attached to these elements (their ranking)
(3) Discover how effective the organisation is in providing these elements compared to competitors
(4) Having collected the data which seek to analyse the organisation's ability to match the expectations of its customers with respect to these elements, consider what changes can be made to implement improvement

Stage one

This stage is difficult because different customers attach varying levels of importance to the service elements they believe they should receive. However, unless some attempt is made to both recognise this fact and, more importantly, to understand what each customer wants, it will be unsurprising if an organisation's customers are dissatisfied. As Christopher and Yallop contend, despite the probability that differences will exist between customers' perceptions of what they believe are important service elements, it is likely there will be some commonality.

The best way to carry out this stage of the model, Christopher and Yallop suggest, is to conduct a small market research type exercise with either the most important customers or a representative sample (where there are many hundreds, for instance). The purpose, they explain, will be to:

- Appreciate the consciousness that the customer has with respect to service elements
- Understand how these service elements compare with other influences such as price, quality of the product/service, availability

On the basis of what is discovered in this stage, it should be possible to collate a list of the elements of customer service for each customer. What is important, they also advise, is to attempt to discover the relative importance that each customer attaches to these elements. Therefore, some form of measure should be included in the survey. This is further developed in stage three.

Stage two

Getting inside the minds of others is never easy; this is what stage two is about. It is inevitable that some customers, when presented with a list of the elements of service that they have identified, and asked to rank their importance, will say that everything is important. This is not helpful to understanding what should be done to enhance their perception of service quality. In order to avoid this tendency, it is possible to use a consumer research technique called 'trade-off'. This technique involves presenting the customer with different combinations of service elements and asking them to provide rankings of these combinations. This technique should provide a useful indication of the perceptions held by different customers of the service they expect to receive. This information is crucial to completing the next two stages.

Stage three

In order for an organisation to gain competitive advantage, it is important to gain some understanding about how well they compare to others who provide products or services that the customer may choose as an alternative. Faced with a choice, we are usually able to make judgements as to who or what provides the best combination of satisfaction. Therefore, the way to find out this information is to ask those who know best about what you provide - that is, your customers. So, using the service elements identified in stage one, it is necessary to ask customers, using a method such as a questionnaire, to rate how effectively they perceive the ability of an organisation to achieve particular service elements when compared to other selected competitors.[43]

The information that this stage provides will allow an organisation to see patterns of how customer satisfaction compares to others in certain segments or geographical areas. The use of a diagrammatic method for presenting results will assist in highlighting these patterns. For instance, Fig. 6.3 below shows the importance that a customer attaches to particular service elements, how well an organisation achieves the provision of these and, finally, how good this organisation is in comparison with a selected competitor.

The virtue of using diagrammatic presentation is that potential areas for improvement should become immediately apparent.

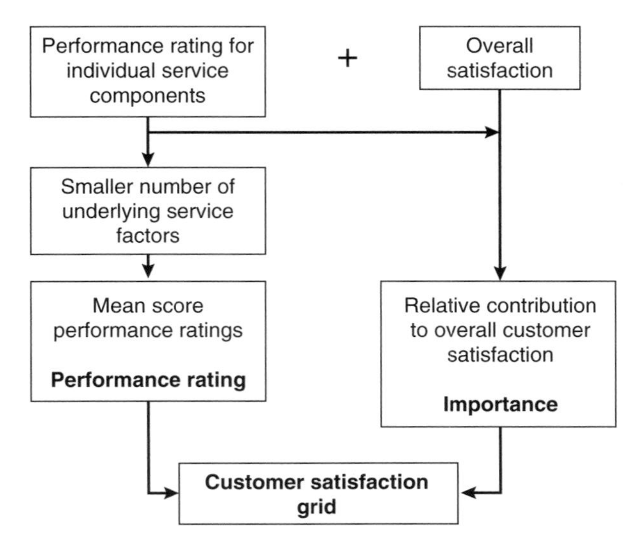

Fig. 6.3 The Prudential model for identifying crucial service factors.

Thus, if such areas emerge, it is a vital part of improvement in customer care to attempt to deal with them. This is what the next part of this process – stage four – seeks to do.

Stage four

This is the ultimate objective of this process – to use the data to implement solutions which ensures that the organisation can do two essential things with respect to customer service:

(1) Seek to maximise customer satisfaction by matching their expectations
(2) Attempt to ensure that what is provided is, *as a minimum*, at least as good as competitors, and as part of addressing long-term improvement, better[44]

6.4.3 The 'Prudential model'

Prudential, like its competitors in the financial advice/insurance sector, is conscious that it is operating in a crowded market. As a consequence, managers are aware that what they give their customers is little different to that which competitors provide. As Prudential managers have come to believe, competitive advantage comes from being *perceived* by its potential customers as giving a better level of service for the same price and product/service than can be purchased elsewhere. In order to do this, Prudential have developed what they call a 'customer satisfaction grid' which, in conjunction with benchmarking against competitors, attempts to derive the importance of service factors, its ability to achieve them and to measure the overall level of satisfaction. Figure 6.3 shows the main features of this model.

In order to effectively use this model, it is necessary to be able to research the following questions:

- What do customers really expect, and what features of service are most critical in terms of giving value?
- What causes customers to move their allegiance?
- Have customers been consulted about what they believe would give them a better experience?
- If competitors are regarded as being leaders in their field, what features of their service provision have allowed them to do this?
- Are there any radical changes that can be implemented to achieve competitive differentiation?

Whilst all of these questions are important, the last - considering radical solutions - is potentially the one that will create competitive advantage. However, whilst being radical has the virtue of being perceived to be very different by *some* customers, there is also the inherent danger of alienating other existing customers. How much an organisation can do in attempting to be radical depends on the context of the service or product and what benefits may be gained compared to any losses. As with all of these models, there is no exact method of predicting the outcomes. What can be guaranteed, however, is that without some effort to understand customers, the likelihood of being 'out of tune' with their expectations will be increased.

6.4.4 The 'efficient consumer response' (ECR) model

This model was originally developed by a number of major companies that either operate in, or supply to, the retail sector in the USA.[45] The companies that developed this model believe that they all have one overriding objective: to provide those customers who buy groceries with the best value possible. As Zairi describes:

> 'The ultimate goal of ECR is a responsive consumer-driven system in which distributors and suppliers work together as business allies to maximise customer satisfaction and minimise cost.' (Zairi, 1996: p. 222)

In order to achieve these objectives, those who proposed the ECR model argued that it is necessary to analyse the entire supply chain and look for ways to ensure that the processes being used are the most effective and efficient. As such, this model has relevance to other industries that operate on the basis of supplier-chain management - such as for instance, the construction industry.

The main principles of ECR are as follows:

- Constant focus on all of the elements that customers perceive as being about value
- Commitment by the organisations involved in the supplier chain (especially senior management) to be committed to the sort of paradigm shift that was described earlier in the chapter
- Using information in a more efficient way to increase customer value
- Intention of value creation at every step (this resembles the main

feature of lean management – see Womack and Jones, 1996 for more information on this subject)
- Common methods of assessing performance

In conjunction with these, there are three elements which need to be engendered in implementing ECR:

(1) Encouraging the need to change attitudes and behaviour
(2) Ensuring those who are involved will 'see it to the end'
(3) Using appropriate technology to transmit crucial information/intelligence as fast and as accurately as possible

In the retail sector, the main focus is that the produce that is being sold provides adequate choice, is fresh, represents value for money, and that the supplier chain through which it travels is sufficient to ensure that the time from origin to shelf is both speedy and cost-effective.[46] Very similarly to TQM, the focus of this method is on the *whole* process, and therefore, involvement of all the participants. Indeed, and consistent with one of the principles of TQM, the emphasis is co-operation of all those involved in the supplier chain to continuously look for potential areas of improvement which will directly result in increased customer satisfaction.

This model has been successfully used to create savings of up to a quarter of the total sales value. What customers experience as a result of shopping at stores that subscribe to the use of ECR is the provision of more choice at a lower cost. Advocates of ECR contend that implementation of this model is possible in industrial sectors other than retailing if there is a sufficiently strong commitment to creating the necessary changes in relationships between all of the 'partners' in the supply chain.

6.4.5 The Conference Board of Canada model

This model originated because of research by The Conference Board of Canada, an organisation dedicated to assist organisations in the implementation of improvement initiatives. Those who were involved in the development of this model analysed the experiences of some of the most customer-focused companies in the world.[47] As a consequence, the principles it contains can be regarded as having been successfully applied by world class performers.

As the researchers discovered by looking at organisations which had demonstrated their ability to continuously satisfy their customers, there are certain vital features which must be in place:

- Integrated management systems which focus on best practice
- Clear objectives which are articulated by the senior management
- An obsession with the use of methods and techniques which cause improvement

In particular, the study identified four essential principles that existed in each of the companies they analysed. These are:

(1) A desire to maximise the potential of every employee
(2) The integration of all effort by every person
(3) Continuous improvement of all aspects of the business
(4) The use of fact in order to manage

Figure 6.4 illustrates the combination of these features and principles.

Whilst this model may seem like a repetition of the principles found in others that have been described, the words 'delighted customer' summarise the main corporate objective of the companies studied. In order to achieve delight, these companies went to great

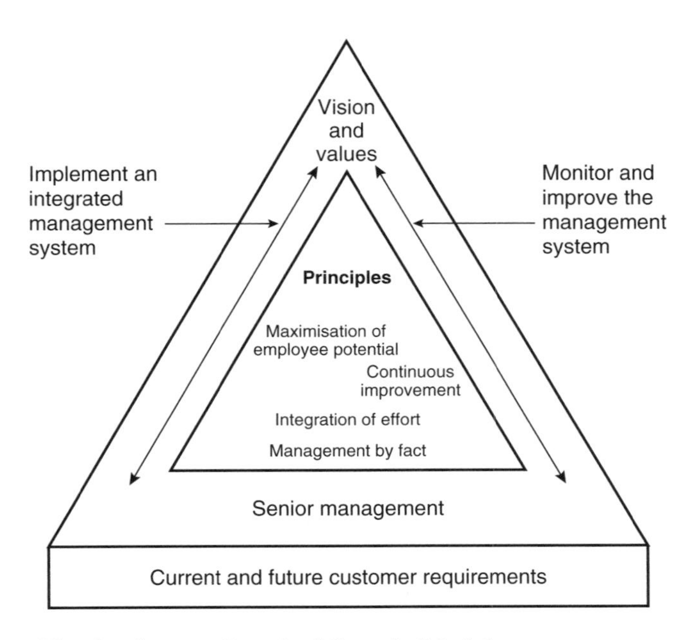

Fig. 6.4 The Conference Board of Canada Model.

lengths to understand the behaviour patterns and concerns of those customers who purchased their products. Moreover, some of these companies, particularly those that were Japanese, were engaged in a search for understanding how society might change in the future. By doing this they believe they could be capable of meeting expectations that might be very different to those which currently exist.[48]

6.5 The experiences of 'best practice' organisations

The following excellent companies tend to have the common feature that they became obsessed with putting customer satisfaction at the heart of their improvement efforts. As such, this section will summarise what methods they use in order to identify and implement methods of enhancing the satisfaction levels of their customers.

Rank Xerox

That this company should feature is no surprise. As previous chapters have described, Rank Xerox is the organisation that tends to be most readily identified with the technique of benchmarking. The need to do this, it should be remembered, was to deal with the fact that their Japanese competitors were producing a superior product which was also cheaper. Rank Xerox realised that in order to survive it needed to use radical measures to increase the satisfaction of its customers. Because of having almost lost its customers, Rank Xerox continually focus on their needs and expectations. As a result, there is a desire on the part of Rank Xerox to ensure that wherever possible, improvements are made to processes which impact on the satisfaction levels of those who buy/use Rank Xerox products.

Since 1984, Rank Xerox has carried out regular surveys of customers by the use of what is called the Customer Satisfaction Measurement System (CSMS). What this involves is shown in Fig. 6.5.

Rank Xerox uses the CSMS to derive feedback that can usefully be applied to develop the operating processes. In order to do this, the surveys focus on two aspects: the products themselves; and the quality of service that is provided. By carrying out this exercise, Rank Xerox believe, they will attempt to ensure that they never again find themselves in a position whereby their survival is seriously questioned. Indeed, so radical has been the transforma-

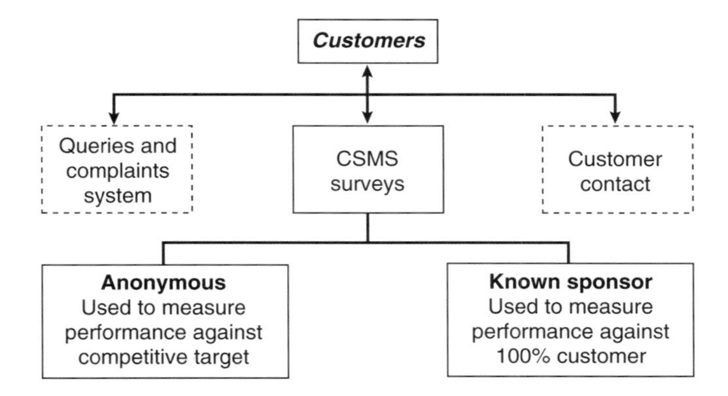

Fig. 6.5 The Rank Xerox CSMS model.

tion of Rank Xerox that it won the prestigious European Quality Award in 1992 (what this involves is described in Chapter 7).

ICL

ICL was able, by using customer care as one of the main business drivers, to create excellence sufficient to be able to be runners up in the European Quality Award in 1993. There are four essential principles of customer care that ICL pursue. These are:

(1) That customer care is perceived to be about exceeding expectations
(2) That in order to create a feeling of being valued by ICL, customers of ICL believe that employees provide an excellent personal service
(3) That ICL are automatically regarded as being the first choice that customers make
(4) Measurement of the satisfaction levels is carried out relentlessly

There are six main methods that ICL use by which to achieve metrics with respect to customer satisfaction. The first three of these focus on data provided by customers of ICL, the other three provide internal data:

(1) *Externally generated*
- Independent surveys
- Score card systems
- Reply cards filled in by customers

(2) *Internally generated*
  - Delivered quality audits
  - Commissioning reports
  - Reports of customer responses to services provided by ICL

Federal Express

Federal Express, winners of the Malcolm Baldrige Award in 1990 (the American quality prize - see Chapter 7), use what are called service quality indicators (SQIs) by which to understand what their customers expect, and to identify areas for improvement. The SQI has a number of components that continually measure the perceptions of service that Federal Express provides to customers. Crucially, the most important indicator that Federal Express analyses for each component, is the *average daily failure point (ADFP)* - something that measures the occurrence of, for instance, non-delivery. The weighting that is assigned to a particular component (non-delivery of items being highest), multiplied by the ADFP, will provide a constant stream of information on trends. Using analysis, these trends can be used to identify and eradicate potential problems. As a consequence, Federal Express - like any customer-focused organisation - will attempt to achieve 100% in every component.

Texas Instruments (TI)

TI won the Malcolm Baldrige Award in 1992 by demonstrating its ability to achieve consistently high levels of customer satisfaction. In order to achieve this, TI uses the following methods:

- Regular customer surveys
- A management tool called the Customer Satisfaction Opportunity System (CSOS)
- The use of what are described as field marketing engineers (FMEs) who are located in, or very close to, the offices of influential customers
- Promotion of communication between senior managers of TI and their counterparts in customer organisations

6.6 Measuring customer satisfaction – what does it all add up to?

The research that the author has carried out into the use of quality management (McCabe & Robertson, 2000) suggests that attempting to *precisely define* what gives a customer satisfaction is extremely difficult. Customer satisfaction involves trying to measure something that, by its very nature, is intangible and therefore will always be a task that seems too nebulous. However, as the descriptions provided in this chapter of how organisations that successfully achieve consistent levels of satisfaction and delight suggest, the need to continually develop a greater understanding of customer expectations is essential. What seems clear is that despite the fact that measuring customer perceptions appears to be a somewhat imprecise science, there are two useful outcomes:

(1) Whatever data are gleaned from the exercise, this will be better than having none at all
(2) The mere fact that an organisation even bothers to ask you what you think will be perceived favourably

Therefore, the need to carry out continual measuring of the satisfaction that existing customers *believe* they have experienced is essential. As the case studies described in this chapter show, this can be achieved by using formal methods such as a questionnaire. This method will provide what is called quantitative data – numbers which can be analysed by the use of statistical techniques. However, as some of the descriptions provided in the case studies suggest, understanding what customers expect requires more than the use of formal methods such as questionnaires; in particular, this will involve engaging in face-to-face discussion. Doing this is commonly described as the 'personal touch'. This allows the person(s) who consume/use the product or service to tell those who are carrying out the interview what *they* (the customers) really think in their own words. Whilst it may be the case that nothing different emerges from carrying out this process, it may be that issues of potential concern are raised. If this happens, it may indicate that certain things have been overlooked. Alternatively, it may be that customers are aware of developments in other sectors (because of their relationships with suppliers and subcontractors) that you might not even have considered. This form of what might be considered 'indirect benchmarking' can be extremely beneficial; showing the ability to respond to customer suggestions will almost always be well-received.

The need for organisations to have a public face is increasingly seen as being important. Many of the most successful organisations in the world can be associated with one person who is seen to personify the characteristics by which the customer strategy of the organisation wishes to operate.[49] In adopting a 'more human' identity, these organisations are attempting to get closer to their customers, and in so doing, to develop a better long-term relationship. As such, these organisations are taking a contemporary approach to achieving what Harold Selfridge suggested was necessary, i.e. learning to please their customers. As the next section describes, a concept called *relationship marketing* exists in order to explain what organisations should do in order to create better relationships with their customers.

6.7 Relationship marketing

In their book *Relationship Marketing* (1991), Christopher *et al.* argue that the concept provides the link between quality, customer service and marketing. As they explain:

> Relationship marketing has as its concern the dual focus of getting and keeping customers. Traditionally, much of the emphasis of marketing has been directed towards the 'getting' of customers rather than the 'keeping' of them. Relationship marketing aims to close the loop. (Christopher *et al.*, 1991: p. 4)

As they also explain, the most important aspect of customer service is the 'building of *bonds*' which will create long-term relationships that allow an 'understanding of what the customer buys and ... how additional value can be added to the product or service being offered' (Christopher *et al.*, 1991: p. 5). There are, they also believe, a number of principles that characterise relationship marketing (Christopher *et al.*, 1991: p. 9). These are:

- Focus on customer retention
- Orientation on product benefits
- Long time-scales
- High customer service emphasis
- High customer contact
- Quality is the concern of all

Relationship marketing has, as its core concept, the need to bring together all of those who are involved in the total process of supplying customers in order for them to achieve more. As such, it is totally resonant with the philosophy of TQM. Moreover, as Christopher *et al.* contend, this concept is differentiated from previous theories of marketing because of the significance that is attached to the influence that employees have when they come into direct contact with customers, 'Ultimately, it is people who develop and achieve competitive advantage' (Christopher *et al.*, 1991: p. 18). Therefore, it is argued, it is essential that organisations that carry out marketing view all employees as having a role to play. Using work carried out by Judd (1987), Christopher *et al.* present a simple matrix (see Fig. 6.6) which categorises people on the basis of how much contact they have with customers and the level of their involvement with what could be called 'traditional marketing'. The import of this matrix is that organisations should more carefully consider the type of person that is most suitable to perform each of the four functions, and the sort of training they must receive.

	Involved with conventional marketing mix	Not directly involved with marketing mix
Frequent or periodic customer contact	**Contractors**	**Modifiers**
Infrequent or no customer contact	**Influencers**	**Isolateds**

Fig. 6.6 People and their influence on customers.

Contractors

A Contractor has most contact with the customer, and will be typically working in a sales or customer services environment. Because of the nature of their job, this person must be fully understanding of the marketing strategy of the organisation. As importantly, they should be prepared and motivated to deal with customers in a way that ensures the perception given is one of a professional and caring organisation. Clearly, when appointing someone to this role, their social skills will be paramount; someone who loses their temper easily will not be best suited. 'Contractors' will, in order to continue to be effective, require regular support and training to ensure that their skills are being enhanced.

Modifiers

A Modifier will be performing a role similar to that of a contractor. Typically, they work as receptionists, telephonists dealing with enquiries and customers' accounts. Thus, even though they may not be working in a 'pure' sales environment, their ability to provide customers with a perception of an organisation that is effective and customer-focused is vital. Like contractors, the sort of person who is appointed to perform a modifier role should be one who will be able to deal with people efficiently and courteously. Also like contractors, they should be continuously trained and developed to ensure their skills in dealing with customers are maximised.

Influencers

Whilst Influencers will have infrequent or no contact with customers, how they behave will have influence on the organisation's strategy for dealing with those who buy its products or services. For instance, an influencer might be located in the distribution department or in researching new developments. What these people do will create lasting impressions with customers. Therefore, their decision-making will require sensitivity and understanding of the implications for customer perceptions. Like contractors and modifiers, continuous training and education to improve their skills to ensure that what they do is dedicated to adding value to customers, is essential.

Isolateds

The Isolateds will have no direct contact with customers. They typically work in purchasing or departments that deal with ancillary functions such as finance or personnel. The fact that Isolateds do not deal directly with customers should not mean that they are considered to be unimportant – as the philosophy of TQM makes explicit, everyone, regardless of their day-to-day activities, can contribute to the improvement effort. TQM, it should be remembered, is about the *total* process. Therefore, every activity that is undertaken will normally be dependent upon other activities; there is mutual interdependence between all activities. If one activity fails or is achieved badly, subsequent activities will inevitably suffer. Therefore, every person – regardless of whether or not they deal directly with the customer – can play their part in improving the final product or service that the customer receives.

6.7.1 'Delighting the customer' – the key to relationship marketing

Relationship marketing, it must be stressed, merely reinforces the message that has been the consistent theme of this chapter: that ensuring effort to retain existing customers will be more sensible than spending time and effort trying to attract new ones. As Christopher *et al.* assert, relationship marketing can be considered as being a ladder (Christopher *et al.*, 1991: p. 22). At the bottom of the ladder are potential customers who must be attracted to buy the product or service in the first instance. Having achieved new customers, it then becomes imperative that an organisation strives to encourage customers to move up the ladder to achieve the following:

- That the experience customers have is sufficient to ensure that, initially, they *support* the organisation by bringing repeat business; and in the longer-term
- That they actively *advocate* others to consider purchasing the same product or service

Any organisation whose customers are predominantly at the top of the 'relationship marketing ladder' will find that the need to spend heavily on advertising is superfluous. The influence of delighted customers will far outweigh any results that advertising can achieve. As those who have been able to build up a network of loyal customers will testify, the need to advertise heavily – a form of expenditure that can entail considerable cost with no guarantee of results – is lessened. This becomes a form of market advantage; if you are able to spend less than competitors it allows more money to be spent on further improvements to the products or services provided.

6.7.2 How to develop a relationship marketing strategy

Figure 6.7 summarises the main elements that are involved in creating the framework for developing a relationship marketing strategy.

Mission

As has been described previously, the first priority for senior management is the need to articulate the main aims, beliefs and

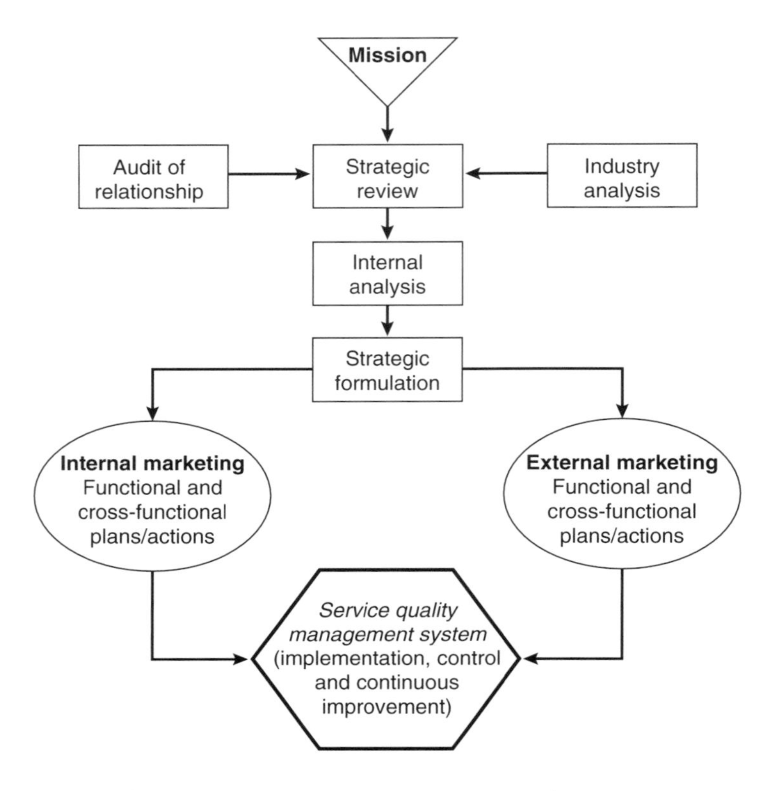

Fig. 6.7 The framework for developing a relationship strategy.

strategies that the organisation stands for.[50] The mission of the organisation provides a vision and focus by which the actions of all employees will occur. According to Christopher *et al.*, in order for a mission statement to be effective, it should achieve the following (Christopher *et al.*, 1991: p. 38):

- Be clear and specific such that no one is in doubt about what the organisation seeks to achieve
- Be focused more on the desire to achieve customer satisfaction than upon technical issues
- 'Reflect the distinctive competences' that the organisation possesses
- Be realistic enough to reflect the strengths, weaknesses, opportunities and threats that the organisation currently faces
- Be capable of being adapted to align the organisation with rapid changes in customer needs

As Christopher *et al.* stress, organisations must continually strive to ensure that what they provide is 'right for the market'. It can be seen that where organisations (such as notable examples in the 'middle customer retail sector'[51]) fail to do this, the end result will be an extremely sudden fall in sales and consequently, profitability. As has been consistently stressed throughout this book, the ability of front-line staff to see potential trends in customer behaviour that may be damaging, will be critical to avoiding such a fate. Therefore, in order to develop a strategy for relationship marketing that is meaningful, these views of such employees should be elicited prior to its creation. Not to do so, risks the danger of creating a strategy that is not only difficult to realise, but worse, could possibly undermine the morale of staff who are vital to long-term development of relationships with customers.

Strategic review and internal analysis

Strategic review and internal analysis occur subsequent to having created a mission statement and having carried out a number of activities that aim to provide accurate indications of the competitive position that the organisation currently enjoys (these are described below). The emphasis of conducting these activities is to achieve, in every possible way, perceptions among customers that the product or service they have received was better than that which could have been purchased from a potential rival. As such, this can be regarded as carrying out what is commonly known as a SWOT (Strengths, Weaknesses, Opportunities and Threats) analysis.

Audit of relationship markets and industry analysis

As the advocates of relationship marketing argue, in order to give the customer 'more' than is provided by competitors, it is essential to make a comparison. Christopher *et al.* suggest that there are five things that have a major effect upon the profitability of any industry, and therefore, more crucially, the ability of individual organisations to achieve particular levels of commercial success. These are:

(1) The difficulties that new entrants face in gaining access
(2) The power that buyers can exercise
(3) The potential for substitution of one product or service for another

(4) The power that suppliers have
(5) The actual level of competition

What is important about these five elements, Christopher *et al.* contend, is that in a market where one or more products/services are in existence, the resulting competition that results from the need to chase customers, frequently means that profit levels are extremely low. Extending arguments that have been propounded by Michael Porter,[52] they suggest that any organisation that is in such a position will find that all of its effort is dedicated to continually having to attract customers (frequently by cutting costs). Alternatively, it is only by shifting the emphasis to one where customers see value in maintaining their custom, that this type of destructive situation can be avoided.

In order to assist in achieving this, they recommend that consideration be given to using what is known as a 'value chain'. This chain seeks to analyse all of the activities that are carried out in such a way as to identify where value is (and is not) sufficiently added. Usually, a value chain is comprised of two types of activity: primary and support. The former are concerned with achieving what must be done in order to create the actual product or service that the customer receives. The latter are activities that provide essential 'props' which allow the primary activities to be achieved. Competitive advantage occurs when an organisation is able to provide greater value than its competitors by carrying out the primary/support activities in a way that is either efficient and/or cheaper, or different in terms of uniqueness. In order to do this, it is essential to allow everyone who is involved in the value chain to identify areas where improvement can be made which will result in customers concluding that the relationship they have is worth maintaining – i.e. they will continue their custom.

Strategy formulation

Having carried out the activities described above, the organisation should be in a position that allows it to consider what market(s) it is best placed to serve, and therefore as a consequence, the most appropriate strategy that must be pursued. This may, of course, mean that the organisation simply concentrates on the market and customers that it currently serves. The organisation should do everything it possibly can to seek ways to add extra value to its customers. However, more proactive action may be necessary in developing a strategy that will be both successful and sustainable.

In order to do this, the organisation should consider one of three strategies:

- Cost-leadership
- Differentiation
- Focus strategy

The first of these - cost-leadership - involves the organisation seeking ways to achieve competitive advantage from being the cheapest. In order to do this, it is necessary to minimise all costs. The second - differentiation - means the organisation must attempt to create competitive advantage by ensuring that its product or service is perceived as superior to alternatives because it offers something very different. What this difference(s) may be, depends on the product or service in question. Examples can include technological features, service or brand image. The advantage to the organisation is that a differentiated product or service may attract a higher price. Thirdly, focus strategy is one that involves the concentration of effort on a particular part of a market, group of sellers or geographical area. This strategy will require the provider to be clear about what potential customers want so as to ensure that satisfaction is achieved.

In deciding which strategy to adopt, managers should be aware of the current position of their organisation in the market, and where they believe it could, by making certain changes, potentially be. Making this decision is, of course, crucial. Case studies of organisations that fail, usually suggest there was either a reluctance to clearly decide on which strategy should be adopted, or alternatively, attempting to achieve a compromise resulted in confusion not only among employees, but more especially, in the mind of customers. The latter, it should be stressed, will usually cause customers to question the wisdom of their continuing to purchase from a particular organisation. As the next section explains, the need to achieve co-ordination between all participants in the overall process is essential in any attempt to achieve better relationships with customers.

Internal marketing and external marketing

Essentially, this part of developing a relationship marketing strategy repeats one of the key principles of TQM: knowing what is most likely to achieve maximum customer satisfaction. Furthermore, it requires the organisation to ensure that high satisfaction levels can be consistently attained. Achieving high levels of customer satisfaction requires effort from everyone. It is essential that each person is aware of what is involved in other activities that constitute the total process (most particularly those activities that occur prior to or subsequent to what they do in carrying out their day-to-day tasks). Achieving this type of approach – the concept of *internal customers* – means that there is an expectation for people to strive to be able to provide others involved in the production or supply chain with exactly what they require in order to give the customer the best possible product or service. Like external customers, carrying out marketing to better understand the expectations of internal customers, is a vital part of improvement.

Christopher *et al.* stress that what customers experience can be considered to be an 'offering' (Christopher *et al.* 1991: p. 57). Using research work carried out by Levitt (1983), they believe that the offering can be viewed on four levels. These levels are:

(1) *Core* – which consists of the physical or intangible characteristics that constitute the product or service. Customers have certain minimum expectations which, in order that they are satisfied, must be achieved.
(2) *Expected* – which consists of aspects of the product or service which customers believe should accompany it. For instance, with most goods, it is expected that there will be a warranty period that covers failures or breakdowns.
(3) *Augmented* – which consists of providing additional features or services that allow the product or service to be perceived as superior to that of competitors.
(4) *Potential* – which consists of the provision of features or elements of the product or service which the customer believes will allow them to derive greater levels of value than could be achieved by purchasing alternatives.

In attempting to achieve a relationship marketing strategy, the focus is on the last two of these levels, but most especially the last. As Japanese producers of electronic and automotive products have demonstrated so successfully, customers who believe they have received products that provide them with high value tend to become both enthusiastic advocates, and more importantly in the long-term, very loyal in terms of repeat buying.

6.8 *Conclusion – the importance of understanding and developing customer loyalty*

The theme of customer loyalty has been one that has been stressed in all of the case studies and models presented in this chapter. Developing loyal customers is the crux of what relationship marketing attempts to create. Therefore, the need to give customers what they want is absolutely crucial. The issues of quality in both the actual product/service and in the service with which it is provided cannot be underestimated. This a point that Peters and Austin make very explicit in their book *A Passion for Excellence* (1985). As they contend, loyalty is established by those companies which are obsessed with understanding what their customers want, and then ensuring that what is provided surpasses these expectations. In order to do this, they believe, it is essential for there to be a robust approach to achieving quality throughout the process, and that whatever systems exist are informed by data elicited from customer perceptions of what they receive.

Buzzel and Gale (1987) have carried out a study of the factors that influence customer perceived quality. In their book *The PIMS Principles: linking strategy to performance,* they assert that an organisation's ability to give its customers more than its competitors can provide will be highly beneficial to any likely return on investment (Buzzel & Gale, 1987: p. 108). Crucially, they conclude, the single most important aspect of business performance is quality when compared to others. If customers are *convinced* they get more from an organisation, they will return to purchase. If they do not feel they are getting the best deal, they will 'vote with their feet'. However, what Buzzel and Gale emphasise is that the method used to ascertain customer satisfaction relies on perception provided by purchasers, not on what managers in the provider organisation *think* will be appropriate or adequate.

The issue of customer satisfaction is one that is again emphasised in the next chapter, which describes the use of excellence awards. As the reader will see, in order that an organisation can win such an award – and therefore justifiably claim to be excellent – it is essential to be able to present data that show how much effort is dedicated to understanding customer satisfaction. Moreover, it is also important to be able to integrate the findings of such surveys into efforts to continually improve the quality of the product/service and the way in which it is provided.

Summary

This chapter has described the paradigm shift that has occurred whereby the customer will no longer accept what they are given; they have much more demanding and sophisticated expectations than would have been the case in the past. As a consequence, organisations that supply goods or service – including, it should be stressed, those that are non-profit-making (i.e. in what has traditionally been known as the public sector) – must, if they want to survive, accept the maxim of the 'customer as king' (Bank, 1992: p. 1).

As a result of the increased expectations of customers, it is essential to be able to understand what they want, and at the very least, to satisfy them. This means an organisation should achieve the following:

- Carry out regular benchmarking of customer satisfaction
- Utilise an appropriate method that allows data to be gleaned from customers, showing how satisfied they are with what you currently provide (a number of models were presented)
- Use the data that was elicited during the customer satisfaction benchmarking exercise to consider ways of creating opportunities for future improvement in both the quality of what is provided, and as Christopher and Yallop's 'surround' model recommends, the way that the product/service is presented
- Appreciate that the so-called 'excellent' companies put obsession with customer delight as the number one strategic objective

CHAPTER SEVEN

HOW TO BECOME WORLD CLASS – THE IMPORTANCE OF MODELS FOR ACHIEVING EXCELLENCE

Objectives

The expression 'world class' has been used frequently in this book. This deliberately reflects the main principle of the concept of benchmarking. It is an expression that is heard with increasing regularity. In sport particularly, the expression 'world class' is commonplace. For instance, in the 1990s, in soccer, Manchester United Football Club achieved domestically a standard of excellence that many accept as the benchmark against which their success must be judged. In this example – which is not atypical of sport – achievement is judged by winning trophies.

The winning of trophies is a useful way to consider how organisations are regarded as world class. As this chapter will explain, the use of what are commonly referred to as 'excellence awards' are, in the contemporary world of business, the accepted way of demonstrating both the aspiration towards, and the actual achievement of, *world-class status*. As will also be described, excellence awards that are used now in Europe can trace their origins to a Japanese award that originated in honour of Dr Deming's influence in the development of quality in the post World War II period.

Specifically, this chapter will describe the following:

- What being a world-class organisation actually involves
- The stages of development that an organisation goes through in order to become world class
- The importance of self-assessment
- The evolution of excellence awards
- What the EFQM Excellence award model is
- How to use the EFQM Excellence award as method to benchmark in order to attempt to achieve world-class status

7.1 What is world class?

For an organisation to be considered as being world class, it is necessary for it to have been conferred with that status. Becoming a world-class organisation is like becoming a world-class sportsman or woman; you have to win prizes to show that you are, literally, the 'best'. Therefore, for an organisation to become world class, it will need to be the winner of one of the awards described below and, as a result, correspond with the following definition:

> [World class] is considered as a point at which a certain standard of practice and performance has been obtained, equalling or surpassing the very best of the international competitors in every area of ... business, such that [an organisation] has achieved international leadership and success. (Blackmon *et al.*, 1999: p. 344)

What is important to note, is that once an organisation has achieved the accolade of being world class, it will be regarded as setting the benchmark against which others measure their performance. Moreover, in order to become world class, an organisation will have demonstrated its ability to satisfy certain criteria contained in the particular award that it will have won. As Chapter 6 stressed, one of the most important things that is considered in such awards is customer satisfaction. Crucially, and as this book has consistently argued, a characteristic that defines world-class organisations is the dedication they put into ensuring that their customers get not just what will merely satisfy them, but that they are delighted by what they receive.

The fact that an organisation has won an award that demonstrates its status as world class will come as no surprise to its customers. As those who buy or consume the products or services of this organisation will have consistently experienced, what they receive will be regarded as being excellent. Importantly, what the customers will experience is a continuing dedication to consistent improvement in the quality of the products and services. As the next section explains, in order to become world class, an organisation must ensure it does certain things, most important of all being its ambition to move beyond the first four levels of the 'UMIST model of adoption of quality management'.

7.2 The UMIST six levels of adoption of quality management

Having studied the use of quality management by European organisations, researchers at the University of Manchester Institute of Science and Technology (UMIST) Quality Management Centre identified the existence of six levels of adoption. These levels are shown in Fig. 7.1. As can be seen from the diagram, the first three levels - uncommitted, drifters and tool-pushers - are on the horizontal part of the curve. This is deliberate. As the labels suggest, these levels of quality management adoption are unlikely to lead to improvement.

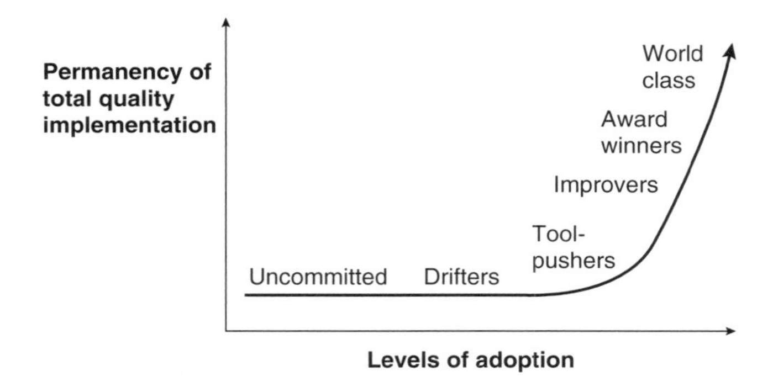

Fig. 7.1 The UMIST model of TQ adoption.

7.2.1 Level one: the uncommitted

The first level, the *uncommitted*, are organisations which, according to Lascelles and Dale, are characterised by very little effort dedicated to 'improvement activities' (Lascelles & Dale, 1993: p. 286). As a consequence, Lascelles and Dale argue, this sort of organisation will find it increasingly difficult to survive.

7.2.2 Level two: drifters

The second level, *drifters*, are organisations which, according to Lascelles and Dale, are characterised by short-term approaches to the implementation of quality management initiatives (Lascelles & Dale, 1993: p. 287). As they suggest, the belief that short-term results can be produced quickly tends to lead to disillusionment, and therefore, a tendency to try something else. The consequence of engaging in this behaviour will, they assert, result in any initiatives being unsuccessful.

7.2.3 Level three: tool-pushers

The third level, *tool-pushers*, are organisations which, according to Lascelles and Dale, are characterised by the use of a range of quality improvement tools that are being used in a 'superficial manner' (Lascelles & Dale, 1993: p. 289). As they contend, this superficiality is caused by a lack of awareness among senior managers that there must be a long-term strategy for attempting improvement. As they explain, until this happens:

> This type of organisation will find it very difficult to sustain the momentum of its improvement initiatives ... [it] gives the right kind of signals, and presents the requisite image to its customers and suppliers, but under the surface is still a 'fire-fighting' culture. (Lascelles & Dale, 1993: p. 291)

7.2.4 Level four: improvers

Therefore, according to Lascelles and Dale, if an organisation wishes to produce change that is likely to be beneficial both to itself, and more particularly to its customers, it must adopt an approach to the implementation of quality management which, at the very least, is characteristic of level four. As a minimum it must attempt to be an *improver*. As they suggest, the senior managers of such an organisation will 'understand that Total Quality involves long-term cultural change and have recognised the importance of continuous quality improvement' (Lascelles & Dale, 1993). Additionally, the following characteristics will be in evidence:

- A 'robust and proactive quality system'
- 'A high degree of closed-loop error prevention'
- Long-term education and training will be commonplace
- Every person will be engaged in personal and organisational improvement activities
- The importance of people is accepted as an essential element of improvement
- There is a strongly committed culture of leadership in existence, and there is at least one 'quality champion' (as Chapter 4 suggests, this person is usually the quality manager)

- 'Benchmarking studies will have been initiated, and the data used to facilitate improvement activities'

However, as Fig. 7.1 indicates, and Lascelles and Dale believe, even though an organisation may have achieved enough to be regarded as an improver, this is not enough. As they argue, there is a danger that the process of improvement may not be 'self-sustaining ... they are still vulnerable to short-term pressures and unexpected difficulties' (Lascelles & Dale, 1993: p. 292) As a consequence of this, they contend, it is necessary for organisations to achieve levels of adoption of quality management consistent with levels five and six.

7.2.5 Level five: award winners

What an award for quality management typically entails is described in a subsequent section (7.6 of Chapter 7). However, for an organisation to be able to even consider itself capable of entering for an award, it will have done certain things that allow it to consider itself as being at least as good as, and potentially better than, any other organisation that has also submitted an application for the same award.[53] What is important to note about this statement is that if an organisation has reached this stage of self-confidence about its abilities, it will also be certain that a 'quality improvement culture' will have become, to quote Lascelles and Dale, 'deep rooted' (Lascelles & Dale, 1993: p. 293). In fact, as they point out, even though winning an award is a very public manifestation of success in achieving excellence, having reached a stage where it considers itself capable, will be a major feat. Lascelles and Dale believe that an organisation that is at this level will have developed the following attributes:

- A culture of leadership that is absolutely committed to improvement in every aspect of the business
- There has been a shift to 'a participative organisational culture'[54]
- There is a commitment from every person to continuous improvement in every aspect of what they do
- Measurement (benchmarking) of the impact of improvement is occurring on a consistent basis
- There is an obsession by every member of the organisation with doing everything possible to 'satisfy and delight customers' (internal and external)

Using a sporting metaphor is a useful way of considering how to evaluate those organisations that are the best. However, as anyone who supports a particular team, individual, or even themselves, will know, not winning first prize can be considered to be failure. Frequently, the cliché 'it's not the winning that is important, it's the taking part that counts' is expressed. Usually, this is of little comfort to those who do not win. However, in the case of attempting to win a quality award, this cliché is true. As will be described in more detail below, when an organisation submits an application to be considered, the process includes feedback by those who assess it as to what can be improved. Clearly, if a subsequent submission is made, this will provide extremely useful information as to how to improve the chances of winning in the future. Moreover, this feedback may truly represent ways to improve the way the organisation carries out its business. As such, this information (which comes from expert assessors) will enable the organisation to become more commercially successful in the future. An example of precisely this situation is quoted by Bruce Woolpert, as President of Granite Rock Company (the first American construction aggregates company to win the Baldrige quality award):

> The Baldrige application process is an investment in future success that every American company should be making. We have applied four consecutive years. Every feedback report that we received contained many suggestions. In fact, one year's report identified 116 areas for improvement. After careful review and discussion, we began working on over 100 of these ideas immediately. Our hard work and improvements were then reflected in the application the following year. (Cited in Porter & Tanner, 1996: p. 252)

This sentiment is precisely what benchmarking involves: the search for continuous improvement which, in this case, comes from those who have assessed an application for a quality award. As the description of the last level of quality adoption indicates, it is to be noted that organisations that are of Japanese origin have, until recently, been dominant in constituting what is regarded to be world class. In such organisations, the search for excellence becomes more than standard practice; it literally becomes an obsession.

7.2.6 Level six: world class

In order to be world class, an organisation must be recognised as being dominant in their field in terms of excellence. As a consequence, these organisations are usually regarded by customers as producing 'the best that one can buy'. Whilst this may mean that their products or services may be more expensive than those provided by competitors, this is not always the case. For instance, in electronics and automotive products, there are producers who consciously serve the luxury, specialist or niche part of markets (examples being, in electronic and automotive markets respectively, Bang and Olufsen, Ferrari and Bentley). If customers could afford the prices that these companies charge, they might be happy to buy 'the best'. Luckily, despite not being able to afford to buy from these companies, people can purchase what are excellent but less expensive alternatives. In the case of electronic products, it is likely that they will have been manufactured by Japanese companies such as Sony or Awai. In the case of cars, even though the vehicle may not actually have been manufactured by a Japanese company, it is widely recognised that the high standards that are now being achieved owe a great deal to the influence of the likes of Toyota and Honda. These companies, like the producers of electronic products, have demonstrated that by applying management techniques which Deming and Juran recommended, it is possible to mass produce cheaply to a standard which, hitherto, had been thought to be impossible.

It should come as no surprise to learn that, traditionally, Japanese companies have tended to be those most associated with being world class. Indeed, Williams and Bersch claimed that in 1989, only ten companies could be regarded as being world class throughout the globe, and all of these were Japanese (Williams & Bersch, 1989). Even though this figure has gone up, the presence of Japanese companies is still notable. As Lascelles and Dale explain, in order to become world class, an organisation will 'be continuously searching to identify more product and/or service factors or characteristics which will increase customer satisfaction through all its networks of process streams' (Lascelles & Dale, 1993: p. 294). As they also explain, such organisations will dedicate their efforts to 'enhancing competitive advantage by [continuously developing] the customer's perception of the company and the attractiveness of the product and service'. This constant dedication to creating excellence in every aspect of what is provided, is called by the Japanese *Miryokuteki Hinshitsu*, which means 'quality that fascinates'. As those who examine Japanese products discover, every aspect of the product and the service that accompanies it will be carried out in such as way as to be better than anything else available.[55]

Whilst it is tempting to think that the potential to become world class only applies to a small number of organisations which operate on a global basis, the features that characterise such organisations can provide inspiration to all. Lascelles and Dale suggest that becoming world class is possible for any organisation that dedicates itself to achieving the following characteristics:

> At this stage the organisation is very close to the goal of TQI [Total Quality Improvement]; where the customer's desires and business goals, growth and strategies are inseparable; where Total Quality is the integrative and self-evident organisational truth; where the vision of the entire organisation is aligned to the voice of the customer. In fact, where Total Quality has become the single constant in a dynamic business environment – it is a way of life, a way of doing business. (Lascelles & Dale, 1993: p. 295)

This approach to doing business is identical to the paradigm shift for customer value strategy that Bounds *et al.* (1994: p. 29) suggest is essential, and which was described in Chapter 6. As Bounds *et al.* also believe, unless organisations are prepared to engage in such behaviour, they will find that there are others who will, and that customers will find the products or services that these organisations provide, far more attractive. As is increasingly seen in the business press, the price of failing to heed this advice can, potentially, be very high.

7.3 *Self-assessment: the key to quality awards*

When quality awards or prizes are discussed, there is a temptation to see the activity as only resulting in the final submission. As has been described previously, whilst winning the award is the ultimate goal, the taking part can be a very valuable part of improvement. As was described by Bruce Woolpert of the Granite Rock Company, many useful suggestions emanated from the assessors of submissions. Usually, for the reason of time and work pressure, many organisations frequently fail to carry out regular reviews of what they do and how they do it. There is also the belief that if you are currently successful, why change 'a winning formula?'. In a successful organisation, this will be particularly so; to quote the popular maxim, 'If it ain't broke, don't try and fix it'. Unfortunately, this sort of attitude can induce complacency. The consequence may be that because of unanticipated changes in consumer behaviour, customers suddenly start buying elsewhere. In short, therefore, if the process of considering how an organisation might be able to comply with the requirements of a quality award achieves nothing else, it should provide an opportunity to carry out evaluation of its existing practices. This, in turn, may give cause for managers to consider changes that may improve the way that day-to-day processes are carried out.

7.4 A short history of the development of quality awards

The first and probably most well-known quality award in the world is the Deming prize. This eponymous award was instituted in 1951 by the Union of Japanese Scientists and Engineers (JUSE) as a way of honouring the work that Deming had carried out in assisting Japanese industry.[56] Whilst it is primarily for Japanese organisations, since 1984 it became open to applications from outside Japan (the first non-Japanese winner being the Florida Power and Light Company in 1989). In particular, its purpose is to recognise the existence of excellence in organisations by the use of what is called 'company-wide quality control' (CWQC).[57] The use of the expression 'quality control' tends to suggest that the Deming prize involves only statistical techniques. As the definition provided in note 57 indicates, the main principles of this prize are broadly aligned with what is known in the West as TQM. As Porter and Tanner explain, enshrined within the Deming prize is the need to ensure that *Kaizen* (the need to continually improve quality) is being applied by all employees and that managers have created a management (control) system that enables them to achieve this objective (Porter & Tanner, 1996: p. 34).

The Deming prize is judged on the basis of an organisation submitting an application which complies with the following ten criteria:

(1) Policies
(2) The organisation and its operations
(3) Education and dissemination
(4) Information gathering, communication and its utilisation
(5) Analysis
(6) Standardisation
(7) Control/management

(8) Quality assurance
(9) Effects (results)
(10) Future plans

Further discussion of the Deming prize is beyond the scope of this book. However, as the following sections describe, this prize influenced the creation in the USA and Europe of equivalents based on similar criteria. Respectively, these are the Malcolm Baldrige National Quality Award (MBNQA) and the European Foundation for Quality Management (EFQM) Excellence Model, (which provides the criteria for the award).

7.5 *The Malcolm Baldrige National Quality Award (MBNQA)*

This prize, like the Deming prize, is named in honour of someone regarded as being influential in the development of improvement in organisations. Malcolm Baldrige was the US Secretary of Commerce from 1981 until his death due to a rodeo accident in 1987, the year that the MBNQA was created as an American equivalent of the Deming prize. Specifically, this award was a direct response to the threat that was being perceived with regard to the quality of Japanese imports into America.

The creation of the MBNQA followed a number of studies – carried out in 1982 and 1983 and sponsored by the American government – into how to attempt to raise productivity and quality to compete with Japanese imports. As these studies concluded (most notably one by the American Productivity and Quality Centre), there was a need to create an award which encouraged organisations to recognise the need to adopt TQM in order to create competitive advantage. As a result, a committee chaired by Malcolm Baldrige was established in 1985 to develop such an award. Following work by this committee, the Malcolm Baldrige National Quality Improvement Act became law on 20 August 1987. The MBNQA encourages the following:

> Understanding of the requirements for performance excellence and competitiveness improvement, and sharing of information of successful performance strategies and the benefits to be derived from using these strategies. (Oakland, 1999: p. 97)

There are, in the MBNQA, seven main criteria and sub-criteria by which assessment of an application is carried out; these are shown in Table 7.1. The way in which these criteria relate to one another is

Table 7.1 The Malcolm Baldrige National Quality Award

Criteria/sub-criteria	Sub-criteria score	Criteria score
1. Leadership		
1.1 Senior executive leadership	45	
1.2 Leadership system and organisation	25	
1.3 Public responsibility and corporate citizenship	20	90
2. Information and analysis		
2.1 Management of information and data	20	
2.2 Competitive comparisons and benchmarking	15	
2.3 Analysis and uses of company-level data	40	75
3. Strategic planning		
3.1 Strategy development	35	
3.2 Strategy deployment	20	55
4. Human resources development and management		
4.1 Human resources planning and evaluation	20	
4.2 High-performance work systems	45	
4.3 Employee education, training and development	50	
4.4 Employee well-being and satisfaction	25	140
5. Process management		
5.1 Design and introduction of quality products and services	40	
5.2 Process management: product and service production and delivery	40	
5.3 Process management: support services	30	
5.4 Management of supplier performance	30	140
6. Business results		
6.1 Product and service quality results	75	
6.2 Company operational and financial results	110	
6.3 Human resource results	35	
6.4 Supplier performance results	30	250
7. Customer focus and satisfaction		
7.1 Customer and market knowledge	30	
7.2 Customer relationship management	30	
7.3 Customer satisfaction determination	30	
7.4 Customer satisfaction results	160	250
Total points		**1000**

shown in Fig. 7.2. As this diagram shows, each criterion is part of one of three elements:

(1) Driver
(2) System
(3) Goals

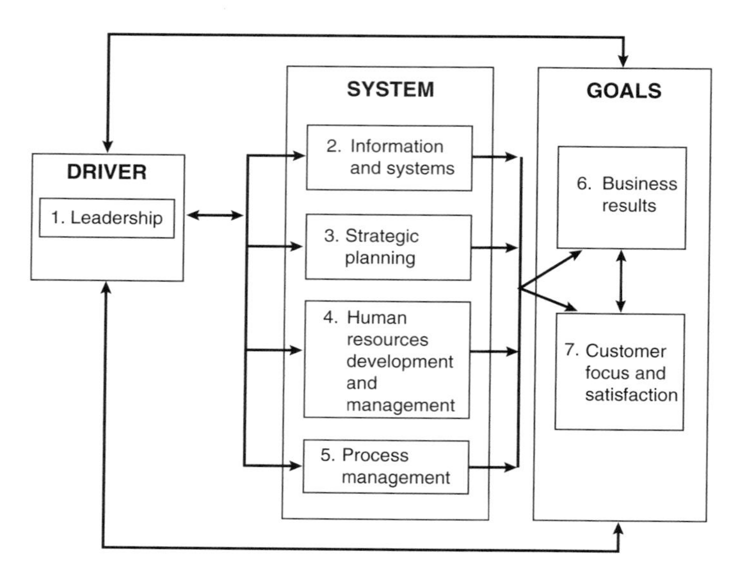

Fig. 7.2 The Baldrige Award criteria framework.

7.5.1 Driver

This element applies to what the managers in the organisation do in order to create direction, vision and objectives that will encourage excellence.

7.5.2 System

This is the assembly of all processes and mechanisms (procedures) that exist in order to ensure the organisation is capable of achieving customer expectations. As such, the use of a recognised quality assurance system would be helpful.

7.5.3 Goals

The aim of the driver and system is to achieve its objectives (goals). These may be of two types:

(1) Customer and marketplace performance - providing what is required at high value and appropriate cost in order to achieve satisfaction
(2) Business performance - the ability to attain all financial and non-financial objectives

As Porter and Tanner explain, these criteria have been developed on the basis of 11 core values and concepts which provide the 'foundation for integrating overall customer and [organisational[58]] operational performance requirements' (Porter & Tanner, 1996: p. 58). These core values and concepts are as follows.

(1) *Customer-driven quality*
This is the effective measure of the customer's perception of the quality of the products or services that they receive. In carrying this out, the organisation should show what they are doing to satisfy existing customers, but also, to proactively identify future expectations.

(2) *Leadership*
Assessment of leadership under MBNQA considers how managers assist in setting strategic direction which ensures that there are 'clear and visible values [which] develop high expectations' (Porter & Tanner, 1996: p. 7). This aspect of the model is, not surprisingly, regarded as being an essential part of creating an appropriate culture for organisational excellence. After all, if managers are unable (or unwilling) to be role models in this process, why should others be prepared to be committed to improvement efforts.[59]

(3) *Continuous improvement and learning*
The need to continuously improve is, of course, the bedrock of using benchmarking. Any organisation using MBNQA would, if the intention was to submit an application for assessment, show how it investigates every aspect of its current methods of operation. The objective is, naturally, to search for ways of carrying them out more efficiently or in a way that ensures greater responsiveness to customer expectations. The word 'learning' was added to this section in 1995 in order to emphasise the need to encourage every person to engage in attempting to learn how the activity they carry out could be improved by finding out how processes are carried out elsewhere.

(4) *Employee participation and development*
If the ability of management to provide appropriate leadership is essential in creating an organisational culture in which excellence will flourish, then the corollary of this must be the way in which people are encouraged to be actively involved in the efforts needed to create improvement. In this section of MBNQA, there is an expectation that support will be provided in the form of training and career development to assist employees to contribute to improvement of the organisation. As a consequence, mechanisms exist to allow people to improve themselves.

(5) *Fast response*
This is something that MBNQA looks for in terms of how the organisation is able to develop new methods of operation, new products or services, and most especially, ability to get these to the customer more quickly.

(6) *Design quality and prevention*
In assessing an organisation under this section, there will be analysis of how well quality and reliability is designed into the product or service as it is being developed. It has been found that the more time that is spent getting this stage right, the less money will be wasted later on attempting to rectify problems later – in terms of both direct costs and customer dissatisfaction.

(7) *Long-range view of the future*
In this section of MBNQA, the emphasis will be on the ability of the organisation's management to anticipate and cope with changes that occur in any context which may impact on the customer's needs or expectations. Organisations that, on the basis of good 'intelligence', are able to predict change will be in a much better position than those who wait until it occurs.

(8) *Management by fact*
This section analyses the basis for decision-making; the assumption being that where robust processes and methods exist, the quality of information will be better. This section of MBNQA resonates with the central message contained in this book, i.e. that if improvement is to be seen to be occurring, its existence must be shown on the basis of objective data, not guesses or hunches.

(9) *Partnership development*
The word 'partner' is one that has become commonplace in the lexicon of business – whether this applies to customers or suppliers. The belief is that where partnerships exist with others, understanding will be enhanced. MBNQA assessment is looking for evidence of the existence of partnerships that are developed with the aim of creating improvement throughout the supplier chain.

(10) *Corporate responsibility and citizenship*
Effectively, this part of MBNQA considers the way that those who manage the organisation consider the impact of the activities undertaken upon all those who may be affected. This responsibility includes those employed directly or indirectly by the organisation – stakeholders. It will also include the local community in terms of residents near offices, factories or sites. If an organisation also shows that it creates links with education, by providing information and assistance, for instance, this would be viewed positively.

(11) *Results orientation*
This section of MBNQA is crucial for every organisation; that what they set out to do, they effectively achieve. For profit-making organisations, achieving adequate profit (or alternatively, return on investment) is usually a prerequisite to continued survival. However, non-profit-making organisations also set targets to be achieved. The ability to achieve anticipated results will be a key measure of the success of an organisation – particularly if this organisation exists to assist certain members of society.

The British Quality Foundation[60] cites findings of research into the use of the MBNQA model by the General Accounting Office in a publication entitled *Management Practices – US Companies Improve Performance through Quality Efforts* (GAC, 1992) which suggests that there are definite benefits:

> high scoring MBNQA applicants had improved their performance in terms of employee relations, productivity, customer satisfaction, market share and profitability. (Cited in BQF, 1998a: p. 35)

7.6 The fundamental concepts of the EFQM Excellence Model

The EFQM Excellence Model is based upon the need for the management of any organisation to dedicate themselves to the task of encouraging and supporting all employees/members[61] to successfully achieve 'eight fundamental concepts of excellence' (EFQM, 1999: p. 4). According to EFQM, in order to achieve excellence, something the Foundation defines as 'outstanding practice' (EFQM, 1999), it is necessary to apply these concepts in a cyclical way (see Fig. 7.3).

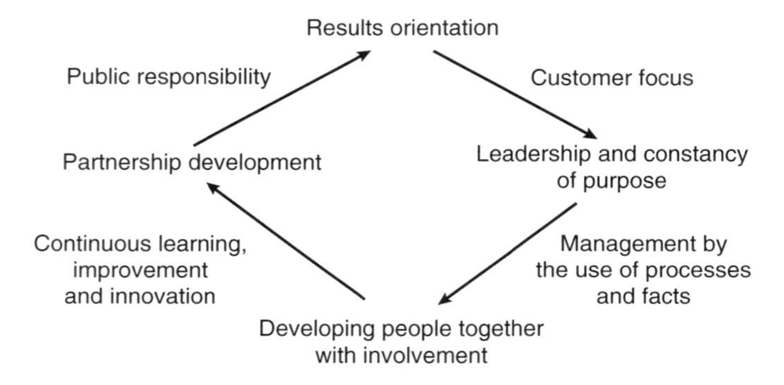

Fig. 7.3 Interrelationship of the eight fundamentals of excellence.

As EFQM advises, the main purpose of using the Excellence Model is for an organisation to understand what it does and, by analysing how it achieves its objectives, create organisational improvement. As subsequent sections of this chapter explain, this model has been used successfully for almost a decade in order to assist in the implementation of excellence in European organisations.

7.6.1 The relationship between regional awards and the EFQM Excellence Model

There is a relationship between the EFQM Excellence Model and British/Regional Quality Awards in that the model is used as the basis for these awards. In the UK, for instance, an award – the UK Quality Award – was launched by the British Quality Foundation (BQF)[62] in 1994. In 1996 the title of this award was changed to the UK Quality Award for Business Excellence, more commonly referred to as the UK Business Excellence Award. In order to recognise that not all potentially excellent organisations feel that they are capable of competing at national level, there are 11 regional Business Excellence Awards. Since 1997, in order to address the different sizes and contexts in which organisations operate, there are separate categories for private and public sector organisations (each of which allows for applications from organisations with more, or with less than, 250 employees). Additionally, since 1998 a category exists for independent small businesses that employ less than 50 employees.

Whilst the likes of the BQF have been successful at promotion of regional quality awards based upon the EFQM Excellence Model, there has been concern by the latter that their intellectual copyright has not been sufficiently recognised.[63] As a consequence, in 1999, in order to ensure that there is no doubt about their origins, EFQM has taken the decision to reassert its intellectual rights over the national and regional variations of their business excellence model. For this reason, any version must now explicitly acknowledge its relationship with the EFQM Excellence Model. Additionally, as EFQM stresses, the word 'business' has been deliberately removed due to the potential perception that non-business organisations might believe it does not apply to them. As such, the ability of EFQM to propagate the Excellence Model to every organisation in Europe is enhanced. In effect, the EFQM Excellence Model can apply to any organisation in Europe, regardless of size, type of operation or nature of business. As subsequent sections will explain, it is an extremely powerful tool for use in benchmarking by organisations.

7.6.2 The EFQM Excellence Model

The original European Quality Model was launched in 1991 as a result of the interest by the business community in a model that would assist in measuring the effects of introducing TQM. As the BQF explains, 'concerns were beginning to surface in Europe in relation to the maintenance of its competitive position within the global economy' (BQF, 1998a). As a consequence, the presidents of 14 leading European companies from both service and manufacturing industries formed the organisation known as the European Foundation for Quality Management (EFQM) in 1989. As those involved in this formation believed, using a prize as recognition of what TQM had achieved in the best organisations could assist others to achieve business excellence. During 1990, a task group from the founder organisations carried out a review of the Deming Prize and MBNQA to consider what lessons could be learned from these models in order to develop a European equivalent.

In October 1991, following the review of existing models for business excellence, the EFQM, with support from the European Union and the European Organisation for Quality (EOQ), officially launched what was called the European Quality Award. This model allows any organisation to assess its ability to achieve excellence. In effect, by doing this on a continuous basis, it is then possible to measure how much improvement has occurred over a specific period. In particular, there are certain key concepts that the EFQM Excellence Model seeks to address. These are:

- Leadership and consistency of purpose
- People development, involvement and satisfaction
- Customer focus
- Supplier partnerships
- Processes and measurement
- Continuous improvement and innovation
- Public responsibility
- Results orientation

As Fig. 7.4 shows – the so-called 'Nine-box Model' – the EFQM Excellence Model is based upon a framework that consists of nine

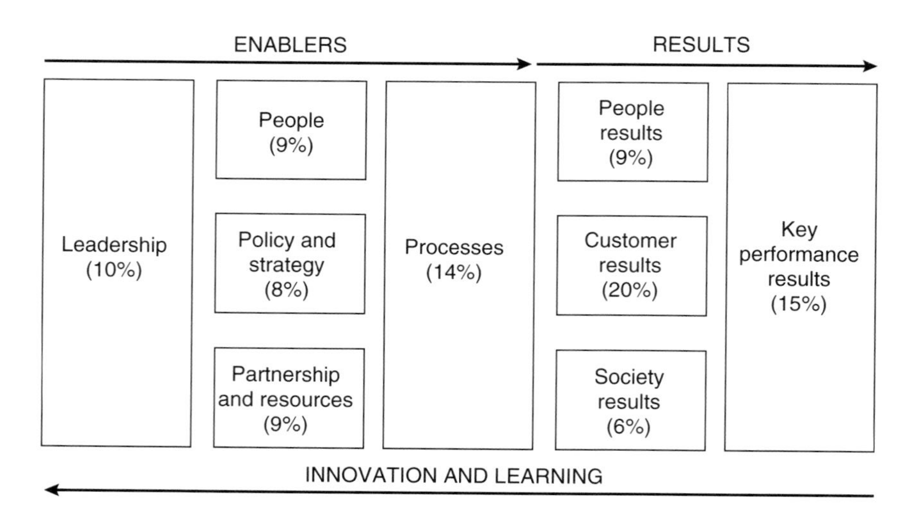

Fig. 7.4 The EFQM Excellence Model®.

criteria (the relative worth of each being shown in brackets). This model, advocates claim, applies to any organisation where the desire of senior management is to achieve the following:

- increase the satisfaction of its customers
- motivate its employees
- ensure that its impact on society is enhanced

As a consequence, advocates of this model claim, the results that this organisation achieves will be improved. The EFQM Excellence Model, therefore, is based upon the principle that, using self-assessment, any organisations should be able to achieve the four *results* criteria through the implementation of the five *enabler* criteria. (What this involves is described in detail below.) Oakland argues that the EFQM Excellence Model can be regarded as a vital part of benchmarking in order to achieve improvement. In particular, he advises that 'systematic review and measurement of operations' is essential in producing improvement (Oakland, 1999: p. 99). As he contends, this model provides the basis upon which the central principle of the contribution of people can be utilised by identifying processes that can be improved:

> The EFQM model [recognises] that processes are the means by which a company or organisation harnesses and releases the talents of its people to produce results performance. Moreover, improvement in the performance can be achieved only by improving the processes by involving the people. (Oakland, 1999)

Porter and Tanner reinforce this message by stating that the 'philosophy of the model is that superior performance is achieved by involving people in improving their processes' (Porter & Tanner, 1996: p. 120). As they explain, in order that people can 'do their best', it is essential – as Deming recommended – that quality is seen by senior management in any organisation as 'a strategic imperative'. As the next section describes, it is therefore no surprise that in the 'Nine-box Model', the first criterion that is seen is that of leadership. As Fig. 7.4 shows, this aspect of the model is what is called an *enabler*. As the next two sections describe, the enabler and results criteria have certain sub-criteria which assist an organisation in understanding the sort of things that the EFQM Excellence Model attempts to assess.

The EFQM Excellence Model enablers

The five criteria of leadership - people, policy and strategy, partnership and resources, and processes - are what are known as enablers; aspects of what the organisation does in pursuit of its overall objectives. They are concerned with *how* an organisation must attempt to do whatever is necessary in pursuance of its objectives. As such, and as will be explained in detail in the section which deals with how an assessment is carried out, enablers are manifested in statements and descriptions which an organisation provides in its application.

In order to provide additional information of the sort of things that an organisation should address, each of the enabler criteria is split into sub-criteria. As well as defining these sub-criteria, EFQM provides indicative suggestions of the things that an assessor will be seeking to find when scoring an application. (What this involves is explained in a subsequent section.) In addition, where they are likely to be considered 'critical' by the EFQM, the book includes descriptions of established business standards or initiatives that might assist an organisation in seeking to create organisational excellence using the EFQM Excellence Model.

Criterion 1

Criterion 1 deals with leadership - how the management of an organisation inspire and lead people to achieve organisational excellence.

Sub-criteria

1(a) Leaders develop the mission, vision and values and are role models of a culture of excellence
2(b) Leaders are personally involved in ensuring the organisation's management system is developed, implemented and continuously improved
1(c) Leaders are involved with customers, partners and representatives of society
1(d) Leaders motivate, support and recognise the organisation's people

Good leadership is, as previous chapters have stressed, an essential part of ensuring that an organisation, or more particularly its people, are being managed in a way that ensures that quality improvement and excellence are accepted parts of the culture. As Porter and Tanner contend, '[criterion one] looks at how the executive team and all the other managers inspire, drive and reflect total quality as the organisation's fundamental process for continuous improvement'. (Porter & Tanner, 1996: p. 124).

As the sub-criteria shown above suggest, there are particular indicators of behaviour that managers in an organisation should demonstrate in pursuance of these objectives. As a consequence, the sort of things which managers should be doing (and therefore described in its written application[64]) would include:

- That there is evidence of all managers being actively involved in the encouragement and support of all efforts to improve the day-to-day processes
- That there is a clear indication that managers recognise what their employees do, and demonstrate recognition
- That managers demonstrate their dedication to improvement by showing how they are prepared to review their own methods of management
- That managers are willing to dedicate, at the very least, sufficient resources to ensure that people are able to implement potential improvement initiatives
- In order to show their commitment to benchmarking, managers are actively involved in carrying out comparison with other organisations. (Being involved in the assessment of organisations which have applied to the EFQM Excellence Model would be seen as an excellent demonstration of this)

A business initiative exists which EFQM considers to be critical in any organisation's attempt to comply with the criterion of leadership: Tomorrow's Company.[65] This is an initiative which seeks to propagate a philosophy that 'advocates an inclusive approach, focused on stakeholder relationships to achieve business improvement' (BQF, 1998a: p. 17). As those who advocate this initiative stress, it is not intended to be a standard like, for instance, ISO 9000 (BSI, 1994) or Investors in People, both of which are described below. What Tomorrow's Company does do, however, is, on the basis of a study which involved 8000 business leaders and influential thinkers, to provide ideas for action which will assist managers of any organisation to create organisational improvement and excellence. As such, it would provide valuable guidance for considering the sort of things that would be required for criterion 1.

Criterion 2

Criterion 2 is about policy and strategy – how the organisation ensures that what is formulated as part of the corporate decision-making process is translated into plans and actions.

Sub-criteria

2(a) Policy and strategy are based on the present and future needs and expectations of the stakeholders
2(b) Policy and strategy are based on information from performance measurement, research, learning and creativity-related activities
2(c) Policy and strategy are developed, reviewed and updated
2(d) Policy and strategy are deployed through a framework of key processes
2(e) Policy and strategy are communicated and implemented

Any organisation, in order to ensure that, at the very least, it survives, and better still, flourishes, needs to have some idea of where it is going, and the objectives it must achieve; this involves the need to engage in planning policy and strategy. As Porter and Tanner explain, '[criterion 2] looks at how the organisation's policy and strategy reflect the concept of total quality and how the principles of total quality are used in the formulation, deployment, review and improvement of policy and strategy'. (Porter & Tanner, 1996: p. 127).

As these sub-criteria would strongly suggest, this criterion concerns the methods and statements that determine the way in which the organisation is managed in order to achieve 'corporate'[66] objectives. As a consequence, developing policy and strategy is an activity that will almost always be carried out by senior managers. The policy and strategy of an organisation should include explanations of the following:

- *Mission* – this is the main purpose or reason that an organisation exists
- *Values* – these are the things that all members of the organisation subscibe to, and would articulate as being central to the organisation's purpose, e.g. commitment to improvement in customer service and relationships, truth and honesty in supplier relationships, support and development of people
- *Vision* – this is the prediction of how senior management wish their organisation to develop in the future

Criterion 3

Criterion 3 concerns people – how the organisation ensures that its key resource (i.e. its employees) is utilised and supported to enable them to contribute fully.

Sub-criteria

3(a) People resources are planned, managed and improved
3(b) People's knowledge and competencies are identified, developed and sustained
3(c) People are involved and empowered
3(d) People and the organisation have a dialogue
3(e) People are rewarded, recognised and cared for

In previous chapters, a great deal of emphasis was placed upon the involvement of people in quality improvement. As Dr Deming always stressed, 'Quality is people not products'. This sentiment is explicitly recognised in this criterion. As the sub-criteria indicate, an organisation considering submission against the EFQM Model should have robust methods of ensuring exactly how people – 'its key resource' – are being 'utilised and supported'. This criterion has a direct relationship with two existing UK awards that seek to achieve exactly this aspiration. These awards are the Investors in People Standard (IIP)[67] and Management Standards – Management Charter Initiative and Vocational Qualifications.

IIP This seeks to encourage organisations to ensure that its employees are developed and supported to their fullest potential and is viewed as an essential part of organisational improvement. As note 67 clearly indicates, criterion 3 and the Investors in People Standard are, to all intents and purposes, identical in what they seek to achieve. For this reason, any organisation that has already complied with IIP should be well-developed with respect to how it manages people. Indeed, the document *Links to the Business Excellence Model* contends that because IIP 'evaluates the investment in training and development to assess achievement against business objectives and improve future effectiveness', it is regarded as being 'critical' (BQF, 1998b: p. 32).

Management Standards – Management Charter Initiative and Vocational Qualifications These are qualifications which are intended to assist managers and supervisors learn what best practice is and, more especially, how they can learn to use such knowledge in their own organisations to create improvement. The aim of using these standards is, according to BQF, 'individual development, corporate gain' (BQF, 1998b: p. 19). In addition, there are four aims of using them:

(1) To create appropriate organisational structures
(2) To use them as a means of analysing skills and, as a result, identify potential areas for improvement in the performance of individual managers or supervisors
(3) To provide a method of benchmarking
(4) To assist in the achievement of other standards described in this chapter (particularly, the BQF stresses, in self-assessing against the EFQM Excellence Award)

Within the Management Standards there are the following roles:

- Manage activities
- Manage resources
- Manage people
- Manage information
- Manage energy
- Manage quality
- Manage projects

Any organisation which intends to use this initiative will, with assistance from their local TEC (Enterprise Company in Scotland), use them to identify organisational or individual management issues that require change. The effectiveness of training or educational programmes to address these problems can be monitored by these standards. The benefit of doing this, according to BQF is that they:

> ... provide a comprehensive and flexible benchmark of best management practice that can be used to improve aspects of an organisation's performance through management development (BQF, 1998b: p. 21).

As such, they provide a very valuable mechanism for an organisation to ensure that it has provided support for its managers and supervisors which will allow it to effectively deal with criterion 3 of the EFQM Excellence Model.

Criterion 4

Criterion 4 deals with partnership and resources – how the organisation has developed and managed its relationship with external parties and effectively used things like finance, buildings, equipment, materials and technology in pursuit of corporate objectives defined in the policy and strategy section.

Sub-criteria

4(a) External partnerships are managed
4(b) Finances are managed
4(c) Buildings, equipment and materials are managed
4(d) Technology is managed
4(e) Information and knowledge are managed

This criterion can be considered to have a direct link to criterion 2 in that it considers how the organisation develops its relationships with partners, and effectively uses its resources to operationalise policy and strategy. As such, an organisation which is considering how it might assess itself against the EFQM Excellence Model must ensure that every aspect of its partnerships and resources have been examined to create continuous improvement.

Criterion 5

Criterion 5 deals with processes – how the organisation shows the way it manages and reviews all processes that are carried out.

Sub-criteria

5(a) Processes are systematically designed and managed
5(b) All processes are improved, as needed, using innovation in order to fully satisfy and generate increasing value for customers and other stakeholders
5(c) Products and services are designed and developed based on customer needs and expectations
5(d) Products and services are produced, delivered and serviced
5(e) Customer relationships are managed and enhanced

Processes are the method by which any activity, regardless of size or importance, is carried out – i.e. the way in which all of the things that should contribute to the desired result are actually achieved. As has been explained in Chapter 5, processes should be considered as a vital aspect of benchmarking, and more especially, how overall improvement can occur.

Like criterion 3 (People), this criterion is considered to be closely aligned to an existing Standard for quality assurance: ISO 9000 (BSI, 1994). Therefore, an organisation that has previously achieved accreditation to ISO 9000 will need to have implemented a quality system that seeks to effectively manage its day-to-day processes. As such, and concomitant with what this criterion seeks to explore, the operation of such a quality management system should be carried out in a way that ensures the key objective of customer satisfaction is consistently attained. Moreover, the quality management system should be operated so as to ensure that improvement is consistently sustained with respect to every aspect of all the day-to-day processes.

The EFQM Excellence Model results

These parts of the Excellence Model, namely criteria 6–9, assess what the organisation has achieved as a consequence of having developed and implemented the enablers which are described above. The results represent the *whats* of achievement. There is a direct connection between the two: results occur because of the successful implementation of the enablers. Enablers tend to be statements about *how* it will seek to achieve certain things. However, results tend to be manifested by more objective numerical data than is possible for enablers. As such, these data can be presented in a submission for assessment that has been included in financial statements or elicited from surveys that have been carried out.

All of the results criteria contained in the Excellence Model have two sub-criteria that are shown below. As for the enablers, descriptions are given of established business standards or initiatives that might assist an organisation in creating organisational excellence using the EFQM Excellence Model.

Criterion 6

Criterion 6 concerns customer results – what an organisation has achieved in terms of satisfaction of its external customers is a consequence of its efforts.

Sub-criteria

6(a) Perception measures
6(b) Performance indicators

This criterion has the highest value attached to it: 20%. This demonstrates the importance that is given to achieving customer satisfaction by EFQM. As has been indicated already, whereas the enabler criteria are about how the attainment of particular objectives are being managed, results need to have objective evidence in order to demonstrate what has actually been achieved. Therefore, and as was stressed in Chapter 6, in order to address customer perception measures, it is vital that an organisation can produce evidence which shows the results of customer surveys or opinions as to the quality of what they receive and how they receive the product/service. Such surveys should seek to elicit how the organisation and, of course, its employees, deal with things like:

- Providing consistency in terms of the product or service
- Delivery times
- Handling of queries or complaints
- Courtesy
- Managing warranties or guarantees

In order to address the performance measures, it will be necessary to show trends of how customer satisfaction has improved over a period of, usually, at least three years. The sorts of things that this would include might be:

- Trends in defects
- Achievement of delivery times
- Complaints received
- Ability to maintain existing customers (repeat business)

There is a business award which EFQM considers to be critical in attempting to address this criterion: the Charter Mark.[69] This award has, like the EFQM Excellence Model, nine criteria against which an organisation attempting to win the award must measure itself; these are:

(1) Service standards and performance
(2) Information and openness
(3) Consultation and choice

(4) Courtesy and helpfulness
(5) Putting things right
(6) Value for money
(7) User satisfaction
(8) Measurable or demonstrable improvements in service
(9) Planned enhancements to services

Clearly, any organisation that is able to win the Charter Mark will be well-placed to deal with the requirements of criterion 6 of the EFQM Excellence Model. As the BQF explains, a number of criteria contained within this award (criteria 3, 4, 5, 7 and 8) specifically address sub-criteria 6(a) of the EFQM Excellence Model (BQF, 1998b: p. 30).

Criterion 7

Criterion 7 is all about people results – what the organisation has achieved with respect to satisfaction of all the people involved in the process of carrying out the activity of providing customers with the product(s) or service(s) that are provided.

Sub-criteria

7(a) Perception measures
7(b) Performance indicators

As the definition of this criterion indicates, the emphasis is on what measures have been carried out in an organisation to show what it has been able to achieve in terms of ensuring the satisfaction of its employees. This criterion is clearly linked to criterion 3 – people. As such, assessors will be looking at how effective the implementation of initiatives described within the sub-criteria of the enabler people has been. Thus, as far as sub-criterion 7(a) is concerned (perception measures), it is incumbent upon the organisation to carry out surveys of its staff to show what their feelings are with respect to things like:

- The provision of opportunities for training and development
- Pay
- Working conditions and environment
- The way they are managed
- The way they are informed about key decisions and strategy
- How they are encouraged to be involved in quality improvement and organisational development

With respect to sub-criterion 7(b), the sort of performance indicators that could be used are:

- The percentage of budget dedicated to training and education
- The provision levels of welfare facilities
- The rates for absenteeism and sickness
- Turnover of employees
- The use of employee suggestions
- Involvement of employees in improvement initiatives

In the description provided for criterion 3 – *people* – it was explained how the use of Investors in People would be valuable. The Investors in People Award will also assist in providing data for criterion 7 (people results), particularly that part of Investors in People which considers how an organisation evaluates the effectiveness of training and development in assisting people to contribute towards improvement.

Criterion 8

Criterion 8 is concerned with society results – what the organisation has achieved with respect to society, and in particular, the communities local to its operations.

Sub-criteria

8(a) Perception measures
8(b) Performance indicators

This criterion considers what an organisation achieves in order to improve and enhance what is referred to as 'society'. This term, it must be accepted, can cover a very wide spectrum. It can include the local community, by the way the organisation provides employment (particularly in deprived areas), its involvement in schools for education, what it does to enhance the local environment by reducing nuisance such as noise or dust, for instance. Because sub-criterion 8(a) is about perception measures, it is necessary for the organisation to have carried out surveys which show exactly what these are, the length of time over which these have been conducted, and as a consequence, any trends that have begun to emerge.

Sub-criterion 8(b) requires the organisation to provide any additional indicators it possesses which show the effect that any initiatives have had on society. These might include:

- Provision of employment
- Its use of 'green' materials
- What it has done to recycle waste
- Reduction in pollution levels
- Data and trends in complaints against the organisation regarding, for instance, noise

There is an international standard that has particular relevance to this criterion. This standard, BS EN ISO 14001 (BSI, 1996), exists in order to provide organisations with guidance on how to implement effective environmental systems. As BQF states:

> The whole purpose of ISO 14001 is to have a positive effect on the organisation's achievement in satisfying the needs and expectations of the community at large. (BQF, 1998b: p. 38)

In particular, the main objective of this standard is to encourage organisations to effectively implement a system that seeks to proactively manage environmental issues. BS EN ISO 14001 has five main headings which, as BQF explains, '[although not requiring] measurement of society's perception directly, ... does require consideration of processes for external communication' (BQF, 1998b):

(1) Environmental policy
(2) Planning
(3) Implementation and operation
(4) Checking and corrective action
(5) Management review

Criterion 9

Criterion 9 deals with key performance results - what the organisation has been able to achieve in terms of financial and non-financial results.

Sub-criteria

9(a) Key performance outcomes
9(b) Key performance indicators

If there were no other reasons why an organisation should consider adopting the EFQM Excellence Model as the means by which to carry out self-assessment, it is the potential benefits that may occur through this criterion which provide the most compelling argument. If using this model is to be advantageous, the one benefit that will be persuasive is a positive change in the 'bottom line'. Therefore, in carrying out self-assessment, it becomes crucial to have evidence that shows how using this model has impacted on its 'key performance results'.

Therefore, sub-criterion 9(a) is concerned with establishing measures of the achievement of the outcomes that the organisation has planned. Whilst these measures may be financial (profit, return on assets, etc.), it is expected that other non-financial outcomes are considered. These may include things like productivity rates, share price, increase in market/segment share and, in particular, repeat business which, it will be assumed, occurs because clients are satisfied with what they have previously received.

Sub-criterion 9(b) is concerned with the use of indicators that are collected over a period of time to show positive trends. Such indicators are, according to EFQM, 'used to monitor, understand, predict and improve these outcomes' (EFQM, 1999: p. 11). Thus, the indicators that could be of assistance in dealing with this part of the model might include defect rates, supplier or subcontractor performance, cycle time from design to completion, ability to maintain existing customers.

The established award that would be of most relevance to this criterion is that which was described as part of criterion 6 (customer satisfaction) – the Charter Mark. Indeed, as BQF contends, the achievement of key performance results is 'critical to success in the Charter mark' (BQF, 1998b: p. 31). Crucially, as BQF advises, criterion 6 of the Charter Mark (value for money) requires that the organisation is able to present evidence of how it has been able to plan and achieve its targets. This, it asserts, will be essential in considering how to deal with sub-criterion 9(a) of the EFQM Excellence Model. As it also advises, criterion parts 1, 2, 5, 8 and 9 of the Charter Mark (respectively these are: Service standards and performance; Information and openness; Putting things right; Measurable or demonstrable improvements in service; and Planned enhancements to services) will provide essential information which will assist in dealing with sub-criterion 9(b) of the EFQM Excellence Model.

7.6.3 How the EFQM Excellence Model criteria are used

In this section, two aspects of self-assessment are considered:

(1) Scoring of the sub-criteria
(2) The difference in approach to scoring of enablers and results criteria

Scoring of the sub-criteria

An organisation which was so excellent that it could not improve any aspect of its processes would receive a perfect score of 100% which accords with the scores shown in the nine-box model. In effect, during the process of scoring, these percentages are multiplied by a factor of 10. Therefore, for instance, leadership has a potential maximum score of 100 points. As was described in the previous section, leadership, like all of the criteria that constitute the EFQM Excellence Model, has a number of sub-criteria. In practice, it is these sub-criteria which are used as the basis for scoring.

In terms of the weighting that is attached to each of the sub-criteria, enablers are very straightforward; whatever the maximum score for the main criterion, it is divided by the number of sub-criteria. So, for example, criterion 1 – leadership – has four sub-criteria. Therefore, each of these sub-criteria can be scored up to a maximum of 25 points. Table 7.2 shows how the maximum scores for each enabler criterion are calculated.

Table 7.2 Calculating maximum scores for enabler critera

Enabler criteria	Overall score	Number of sub-criteria	Score for each sub-criterion
Leadership	100	4	25
Policy and strategy	80	5	16
People	90	5	18
Partnership and resources	90	5	18
Processes	140	5	28
Total score for enablers	500		

The results criteria of the EFQM Excellence Model are not like the enablers, in that they are equally scored (the only exception to this is criterion 9 – results). As Table 7.3 below shows, sub-criteria (a) and (b) of customer, people and society constitute different percentages of the total score.

The different percentage scoring between what are either perception measures (sub-criteria (a)) or performance indicators

Table 7.3

Results criteria	Overall score	Number of sub-criteria	Score for each sub-criterion
Customer results	200	2	(a) 150 (75%) (b) 50 (25%)
People results	90	2	(a) 67.5 (75%) (b) 22.5 (25%)
Society results	60	2	(a) 45 (75%) (b) 15 (25%)
Key performance results	150	2	(a) 75 (50%) (b) 75 (50%)
Total score for results	500		

(sub-criteria (b)) gives a clear indication of the importance that is attached to particular parts of the model.[70]

The difference in approach to scoring of enablers and results criteria

Enablers The enablers criteria of the EFQM Excellence Model represent statements of how these aspects are carried out. When scoring an enabler, it is necessary to consider two things:

(1) How excellent is the approach that is used for a particular enabler criterion (what is called *approach*)?
   (a) Is it appropriate?
   (b) Is it based upon sound judgement?
   (c) Is it flexible and capable of being adapted to meet changing circumstances?
   (d) Does it relate well to other enablers?

(2) How widespread is the actual use of the enabler in the organisation (what is called *deployment*)?
   (a) How effectively is it being used at the levels of the organisation to which it applies?
   (b) What evidence is there of this effectiveness?

When scoring the sub-criteria of an enabler, it is usual to consider approach and deployment as being two separate parts of the same thing. As the example below shows, the overall score that is given to a sub-criterion is a judgement of the effectiveness of the combination (called *consensus*). In order to assist in carrying out scoring of approach and deployment, EFQM provide five different percentages that could be applied to the score of each sub-criterion (0, 25, 50, 75 and 100%). Thus, 0% represents almost non-existent approach or deployment, and 100% represents the maximum. The scores between 0 and 100% represent increasingly better levels of achievement (descriptions of which can be found in relevant EFQM publications, such as EFQM, 1999).[71] It is possible to use percentages that are more accurate than these 25% intervals. However, the need to achieve such accuracy is not essential; carrying out consensus allows such intervals to be a reasonable indication of achievement (particularly when there is a large difference between the score given to approach and deployment).

Thus, for example, if the sub-criterion of any enabler had the following scores:

Approach 50%[72]
Deployment 25%[73]

the simplest thing would be to apply an average of 37.5% to the maximum score that can be attained for this sub-criterion. Whilst this will be perfectly acceptable, the assessor may feel that, even though deployment of this sub-criterion could be better, the effort that has been put into the approach means that the score should be closer to the 50%. Alternatively, it may be that the assessor believes that the score given to the deployment is a realistic reflection of how well this sub-criterion is being applied and, therefore, it should be scored closer to 25% than 50%. As should be clear from this process, much depends upon the skill and experience of the assessor.[74] It is worth pointing out that this is not 'rocket science'. Remember, the highest score that can be attained for the sub-criterion of any of the enabler criteria is 28 points (processes). Therefore, the consequence of a 10% variation in the scoring of any of the process sub-criteria will result in a change in its total score of no more than 2.8 (not particularly significant when the total potential score is 1000 points).

Results The results criteria of the EFQM Excellence Model represent what the organisation has, by using the enablers, achieved, or is achieving. When scoring results criteria, it is necessary to consider two things:

(1) The level of achievement of the results that the organisation has attained (the *results*)

   (a) How good are they compared to its targets?
   (b) What industry comparisons are used in order to set the targets?
   (c) The length over which the results have been collected

(2) The extent of the organisation or processes to which the results apply (the *scope*)
   (a) How much of the organisation do the results apply to?
   (b) What processes are covered by the results?

The process that is used for scoring of the sub-criteria of the results is the same as that used to score the sub-criteria of the enablers. Thus, the results and their scope are scored separately using, for instance, the EFQM guidance notes of what would be expected for 0, 25, 50, 75 and 100%. Like scoring the enablers, the process is one that is dependent upon the subjective judgement of the assessor. Similarly, the overall score that is applied to the sub-criterion of a result is a matter of making a judgement as to what is an accurate reflection of how well the organisation is achieving it. For example, consider the following scores for the sub-criterion of a result:

Result 75%
Scope 25%

The most appropriate score that should be given to this sub-criterion is a matter of considering what the actual result is, and how much it applies to the processes for which is it relevant. However, in this example, whilst the result attracts a high score,[75] the low score that the scope is given[76] suggests that it could be more widely applied than is presently the case. This would suggest that the overall score for this sub-criterion should, unless there is a compelling reason to do otherwise, be closer to the 25% than 75%. The principle of carrying out scoring is that the assessor can only consider what information is available. If this task is a self-assessment (internal), it is possible for the person to either know or request additional information. However, if an assessment is to be carried out by someone external to the organisation[77] the need to present information in as concise and unambiguous a way as possible is crucial. As the next section describes, the need to use a method which will create scores that are consistent with what external assessors expect, is very important; such a method – RADAR – exists to do this.

The use of RADAR® and Pathfinder to consider EFQM excellence criteria

RADAR® is an acronym that stands for Results, Approach, Deployment, Assessment and Review. As EFQM advises, the use of scores, particularly where they can be compared to organisations that are regarded as being the best, can be a very effective way of carrying out benchmarking in order to compare areas identified as being strengths or capable of being improved. Because of the desire to be compared to the 'best', an organisation that uses RADAR is likely to be very mature in its use of the EFQM Excellence Model. As is described below, if an organisation wishes to use a method of assessing itself against the relevant criteria, but believes RADAR to be too advanced, it can consider what is called Pathfinder.

As EFQM recommend, the use of the RADAR Matrix is a method useful for scoring by considering the following:

(1) *Results:* Do they form part of the organisation's policy and strategy-making process. As such, the following aspects will be considered:
 (a) Trends
 (b) Targets
 (c) Comparisons
  Causes
  Scope

(2) *Approaches:* Are they being implemented in such a way as to produce the required results. As such, are they:
 (a) Sound?
 (b) Integrated?

(3) *Deployment:* Are the approaches being implemented in such a way as to ensure that the desired results will occur. As such, are they:
 (a) Implemented?
 (b) Systematic?

(4) *Assessment and review:* Is the organisation carrying out the above in a way that shows there is dedication to learning from experiences and continuous improvement. As such, the following aspects should be considered:
 (a) Measurement
 (b) Learning
 (c) Improvement

Pathfinder is a method that seeks to find the strengths and areas for improvement that exist in an organisation, without the need to use RADAR. The objective of this method is to assist self-assessment by the use of simple questions. As EFQM explains, whilst Pathfinder:

> ... reflects the RADAR logic ..., it is not a prescriptive list, rather it offers guidance to organisations on what steps need to be considered on the route to excellence. (EFQM, 1998b: p. 30)

Pathfinder, therefore, considers the following aspects:

(1) *Results*
   (a) Do they apply to all relevant parts of the organisation?
   (b) Are both perception and indicators included?
   (c) Do they show consistent improvement over a sustained period of time?
   (d) Are they compared to pre-set targets?
   (e) How well do they compare to industry averages?
   (f) Are they linked to the implementation of enablers?

(2) *Approaches*
   (a) Are these based on sound logic?
   (b) Do they support the implementation of policy and strategy?
   (c) Are they appropriate?
   (d) Is there a link with other approaches?
   (e) Can they be sustained?

(3) *Deployment of the approaches*
   (a) Widely across the organisation?
   (b) Used fully?
   (c) In a systematic way?
   (d) Can be measured?
   (e) Easy to understand?

(4) *Assessment and Review of approaches and their deployment:*
   (a) Is this carried out regularly?
   (b) Is learning being carried out
   (c) Is comparison (benchmarking) being achieved?
   (d) Is continuous improvement occurring?

7.6.4 What is considered when using the EFQM Excellence Model to benchmark

As EFQM explains, the main benefit of using the Excellence Model is that it will be beneficial in assisting an organisation to think about how it should develop itself:

> When looking to the future, developing strategies and aligning the organisation to deliver those strategies, it is vital for the organisation to understand its present strengths and its weaknesses, or areas for improvement. [In order to do this...] a process the European Foundation for Quality Management (EFQM) calls Self-Assessment, provides enormous insight. (EFQM, 1999: p. 3)

As EFQM contends, its model provides a 'large-scale view' of the organisation which will allow the management team to identify what are the most important aspects of its business that require improvement. Moreover, EFQM argues, its model, because it 'includes the dynamic of benchmarking against world class organisational practice', enables the management of any organisation to consider how it might create opportunities to become similarly regarded. The message that EFQM is keen to stress, is that once an organisation has embarked upon the use of its model, provided this is done in the 'correct spirit', it is inevitable that improvement should occur. As this section describes, the main objective of self-assessment is to analyse best practice that exists in the organisation (referred to as *strengths*), and perhaps more importantly, in order to develop the organisation's capabilities, any potential weaknesses that will undermine its abilities (referred to as *areas for improvement*).

The EFQM recommended approaches to self-assessment

A number of approaches exist to assist organisations to use self-assessment as a method to improve, the appropriateness of which will depend upon how mature its efforts are in achieving excellence using the model, and the amount of time and effort it wishes to input to the task. The list below contains the full range of approaches that EFQM recommends in an order which shows increasing complexity and relevance to organisations that are well-developed in their implementation of the excellence model:

- Standard matrix
- Elementary questionnaire
- Standard questionnaire
- Very detailed questionnaire
- Tailored matrix
- Questionnaire and workshop
- Matrix and workshop
- Pro-forma
- Facilitated workshop

- Pilot award simulation
- Pro-forma and workshop
- Appropriate questionnaire
- Pro-forma supported by peer variation
- Award simulation

The decision about which of these is most appropriate is dependent upon the maturity of the implementation of improvement in an organisation. When the management of an organisation consider the EFQM Excellence Model for the first time, they may possibly be daunted by what it contains (the nine main criteria and their 32 sub-criteria). The biggest concern may be how it is possible to dedicate the additional time and effort that using this model requires. As EFQM suggests, there are ways of carrying out self-assessment which, whilst they are impossible to carry out without at least some additional effort, can assist the organisation to learn the following:

- What its strengths and areas for improvement are
- How best practice can be applied (either that which exists within the organisation or from other organisations)
- How much needs to be done in order to create excellence
- Where additional effort should be dedicated (particularly with respect to supporting the development of people)

Regardless of whatever approach is used, the way that the exercise is conducted will follow the pattern shown in Fig. 7.5. The list of approaches shown in the diagram can be categorised into five main approaches for self-assessment. As the explanation provided for these suggests, the choice of which is most appropriate for an organisation depends on the time and effort it believes it can dedicate.

Approach 1 – questionnaire

The questionnaire is the simplest method of carrying out self-assessment. In essence, it relies on the use of standard questionnaires that EFQM has prepared by which to gather data on each of the nine criteria and their sub-criteria. As EFQM admits, whilst this approach has the benefit of being quick and easy to use, it will be dependent upon the following factors:

- It may not be clear what the areas of strength and improvement are

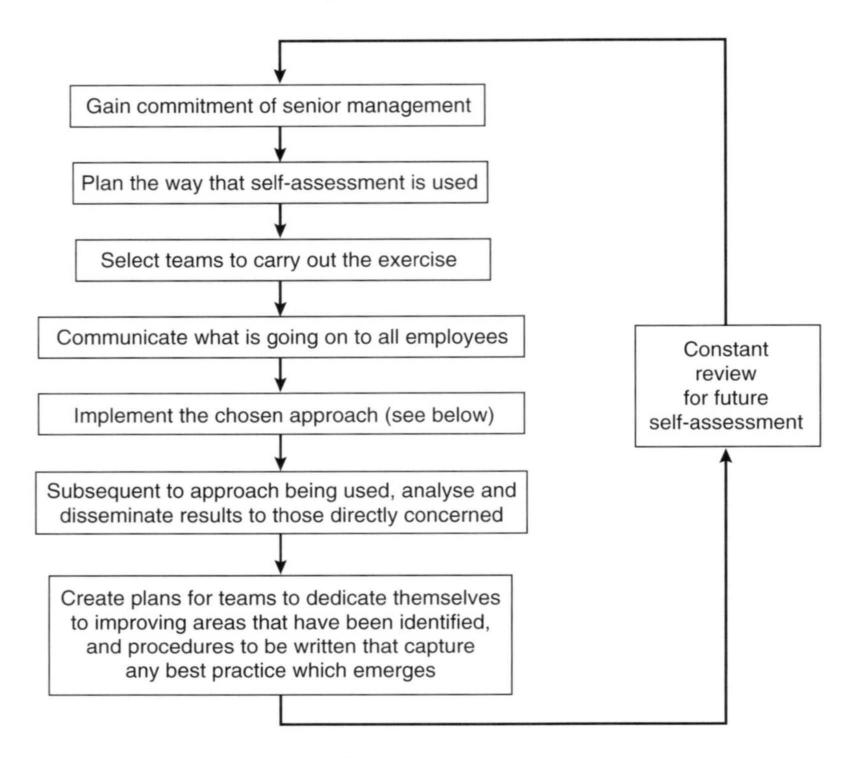

Fig. 7.5 The overall approach to self-assessment.

- The data may be superficial
- This form of data collection does not always attract a particularly high response rate
- The answers may beg more questions than they solve
- Scores will not emerge for the criteria (comparison with other organisations is therefore more difficult)

Approach 2 – matrix

In using the matrix method of self-assessment, the objective is for the organisation to create ten of what EFQM calls 'statements of achievement'. The creation of these statements will be on the basis of what are found to be realistic levels of achievement of particular criteria.[78] The ten statements are graduated upwards to indicate increasing levels of achievement (one is poor, ten is excellent).

The matrix is then used to assist in the operation of a workshop which involves the following four steps:

(1) *Briefing.* To introduce those present to the aims and objectives of the exercise and, in particular, to demonstrate how the scoring works (to allow people to carry out individual rating).
(2) *Individual rating.* This occurs subsequent to the briefing, and requires those who are involved to use the matrix and decide which of the statements provided most accurately reflect current practice.
(3) *Consensus meeting.* This meeting is organised once those involved in the exercise have carried out their rating of the statements contained in the matrix. The objective of this meeting is for someone (preferably a trained facilitator of the EFQM Excellence Model) to ensure that all those who have carried out the exercise can come to 'broad' agreement.
(4) *Action planning.* This can be considered to be the 'pay-off' from having carried out this exercise. If, having carried out exploration of what people think actually goes on, areas are identified as needing to be improved, how this may be achieved will need to be planned so that adequate resources and attention are provided by those most intimately involved.

The use of a simple matrix has a number of advantages, most especially that it can be easily understood and used by employees regardless of hierarchical level. As is described in the steps above, the real value is in getting people to discuss their perceptions of what does (or does not) happen. As a consequence of these discussions, issues may then emerge which will allow the identification of potential areas for improvement. However, whilst the identification of these areas will be useful, they will not be in the same format to allow direct comparison with the EFQM Model proper. If such comparisons are required, the alternative methods, as described below, can be considered.

Approach 3 – workshop

As the title suggests, the workshop approach involves the assembly of a number of people who have direct involvement in the processes that particular criteria contained in the EFQM Excellence Model rely on. As EFQM recommends, this method is one that should involve at least two trained EFQM assessors, one of whom is from the organisation, and the other who is external (and therefore capable of being objective). This approach requires those involved to collect data in such a way as to allow the management team of the organisation to 'reach consensus' (EFQM, 1999: p. 20). A workshop approach to self-assessment involves the following five steps:

(1) *Training (prior to workshop proper).* This is an essential part of running the workshop: the need to ensure that all those involved are fully acquainted with how the EFQM Excellence Model operates and, in particular, how the scoring system is used with respect to the strengths and areas for improvement of the nine criteria and 32 sub-criteria. As EFQM suggests, a day should be dedicated to allowing those attending to read all the material and carry out a simulation of a pre-prepared case study. Having achieved this, members of the team involved in the workshop should be competent to start the next stage – gathering data.

(2) *Data gathering.* It is important that those who are involved in the workshop should carefully consider their ideas and beliefs about what are the particular strengths and areas for improvement for all of the criteria and sub-criteria. This data will be vital to carry out the next stage – scoring of the sub-criteria.

(3) *Scoring workshop.* As should be apparent from the description of what the model contains (32 sub-criteria), the amount of time that this stage requires is not insignificant. However, if those involved have been briefed to come prepared with the necessary information, time can be used effectively. Whilst it is advisable to put a limit on the amount of time this task requires, it is important to allow adequate time for discussion to take place. This discussion will assist in allowing consensus to emerge about the strengths, areas for improvement and the scores that should be allocated to all of the sub-criteria.

(4) *Agreeing areas to improve.* The logical step that should follow the scoring workshop is to consider how any strengths or areas for improvement that were identified during preceding stages of this process can be used for the purpose of benchmarking or improvement of the organisation. It will be very useful if those who were involved in the workshop are prepared to be responsible for taking ownership of such issues that emanate from the scoring workshop.

(5) *Review of the progress.* As part of any process of continuous improvement, it is vital that there is regular review of the progress being achieved. This stage will allow for regular reporting and, most especially, issues of learning to be shared as part of the process.

This approach to self-assessment is more detailed than the previous two. In particular, such team-building which requires the need to bring different members of the organisation together can provide a valuable opportunity to discuss issues that they may not have previously been able to raise. Doing this, EFQM suggests, should allow 'discussion and agreement ... on the strengths and areas for improvement to build a common view' (EFQM, 1999: p. 21). However, it also warns that this method does have certain limitations:

- It is not as detailed as the two remaining approaches described below
- There is a serious risk that if those involved do not fully contribute, the workshop will produce less than is anticipated
- The data/scoring is based solely upon those involved in the workshop. (Does what they believe truly represent every aspect of what goes on in the organisation?)

Approach 4 – Pro-forma

The pro-forma approach is a method of carrying out a self-assessment which, despite being almost like an award simulation (see below), is not quite as demanding. In order to use this method, it is necessary to create pro-formas that apply to all of the 32 sub-criteria which constitute the EFQM Excellence Model. Each pro-forma is used to elicit knowledge from people[79] in the organisation on the following:

- What are the perceived strengths of this sub-criteria?
- What are the perceived areas for improvement of this sub-criterion
- Does evidence exist to substantiate any of the strengths that have been identified?

The layout of the pro-forma will depend on what information is being collected. However, the principle that simplicity and clarity are paramount should be adhered to.

Whilst it is perfectly acceptable to allow employees to complete the pro-formas individually (in order to collect a great deal of information quickly), it is desirable that there is provision for people to meet as teams. These meetings, it is anticipated, would be used to attempt to ensure consensus. The need to ensure consensus is important if scoring is to be carried out.

Once this method has been carried out internally, it is relatively easy for the output of the exercise to be analysed externally by, for instance, a person(s) who is accredited to carry out assessment in accordance with the EFQM Excellence Model. Additionally, if external assessors are used, it is desirable that they carry out follow-up sessions, the purpose of which is to explore how it may be possible to carry out actions which will create improvement in those areas identified as being necessary.

Once this method has been used once, it provides a benchmark against which its subsequent use can be compared. Clearly, any best practice in particular areas/offices/divisions can be applied elsewhere. Equally importantly, any areas that are considered to be capable of being improved should be regularly reviewed in order to demonstrate that progress is being made.

Approach 5 – award simulation

Because this method of self-assessment is one that involves attempting to simulate what is, in effect, a submission which accords with the EFQM Excellence Model, it is the one which requires most work. However, if an organisation is actively considering submitting an application within a couple of years, carrying out an award simulation is probably an excellent way to prepare.

This method requires the organisation to consider every aspect of the EFQM Excellence Model and, as a result, it will be necessary to do the following:

- Appoint a team of managers/supervisors who are to be responsible for the completion of various tasks
- Ensure that those who are involved are appropriately trained in the use of the EFQM Excellence Model
- Ensure that adequate time and resources[80] are allocated to the task of data collection, analysis and writing up of the final report. It is worth noting that this report, which is submitted by any organisation to the awarding body carrying out the assessment, has a maximum limit of 75 pages for large organisations (35 for SMEs (small and medium enterprises))

One of the main advantages of carrying out an award simulation is that it allows the management of the organisation to appreciate where resources should be dedicated if it is to submit a submission for an excellence award proper. As such, and as is consistent with the philosophy of the EFQM Excellence Model, this is a valuable opportunity to address a wealth of issues that impact upon the organisation's ability to create improvement through both its processes and people.

As in the case of the pro-formas, in order to provide the potential for additional value-adding, the use of an external assessment would be highly advantageous. This would provide an opportunity for 'fresh eyes' to consider the strengths and areas for improvement that are proposed and, in particular, the scores that have been allocated to each of the sub-criteria (and therefore, the overall score that the organisation believes it is capable of). Because the use of scores is an essential part of evaluating applications for the EFQM Excellence Model, if an external assessor is used, it is vital that he or she has recent experience of carrying out evaluation of applications. Such a person(s) can assist in identifying methods of carrying out benchmarking of any sub-criteria or process against those of another organisation(s) that is regarded as being the 'best'. This, of course, is something that should be considered as being a vital part of the process of continuous improvement.

Whilst carrying out an award simulation has the obvious advantage of providing an excellent opportunity for evaluating the organisation's abilities and benchmarking, it is important that those who are involved realise that this exercise - particularly when carried out for the first time - is the first step in long-term continuous improvement. As may be a temptation in the first instance, the tendency to apply scores that are either too generous or miserly should be discouraged; the aim is honesty. High scores may cause complacency, not something that benchmarking is intended to produce. On the contrary, low scores (despite being a disappointment) will clearly indicate where effort should be dedicated in the future. Whatever emerges, as has been stressed in previous chapters, the importance of people's co-operation and support cannot be overstated. Consequently, anything that seems like a 'witch hunt' or excuse to blame people for past failures will be doomed from the start.

Using the output of self-assessment which arises from use of the EFQM Excellence Model

Table 7.4, which is based upon information provided by EFQM (1999: p. 26) summarises the virtues and deficiencies of the five methods for carrying out self-assessment.

As EFQM recommends, whilst each has certain advantages or

Table 7.4 Relative merits of various self-assessment approaches.

	Strengths and AFIs	**Accuracy of score**	**Knowledge of model**	**Site visits needed**	**Trained assessor**	**Resource implications**
Questionnaire	No	Low	No	No	No	Low–medium
Matrix	No	Low	No	No	No	Low–medium
Workshop	Yes	Medium	Yes	No	Yes	Medium
Pro-forma	Yes	Medium–high	Yes	No	Yes	Medium
Award	Yes	High	Yes	Yes	Yes	High

disadvantages, it may be advantageous to consider using a mixture of methods for self-assessment. The objective is for the organisation, in whatever way possible, to create improvement. However, as EFQM stresses, 'self-assessment will not improve your organisation [this can only be] done by people' (EFQM, 1999: p. 33). In order to do this, it is essential that action plans are developed and implemented by employees who have been trained to be capable and motivated. One of the potential problems of carrying out self-assessment using the EFQM Excellence Model is that people begin to feel overwhelmed by the desire of some to improve everything immediately. This is a situation in which management of this process is crucial. As EFQM explains:

> Organisations are unlikely to have the resources to address all these opportunities concurrently and it would be unrealistic for them to try. (EFQM, 1999: p. 36)

Therefore, it is essential that those who are responsible for the task of applying the principles of the EFQM Excellence Model ensure that there is a way to prioritise what problems should be tackled first; something that EFQM calls 'the vital few'. As EFQM advises, what is crucial is to decide how any priorities can be linked to the organisation's key business objectives, the most important of which is customer satisfaction. Clearly, the decision-making process about such priorities must consider what can be achieved given existing levels of resources or expertise. EFQM provides the following five steps to ensuring that the output of self-assessment is effectively used.

(1) *Step one.* Collation of the strengths and areas for improvement in a way that ensures there is a logic and rationale for action. This should be done using information that is coherent and as soon after collection as possible.

(2) *Step two.* Decide upon the method and criteria to be used in order to prioritise the output of the self-assessment. In doing this, their impact on the following should be borne in mind:
   (a) Strategic intentions
   (b) Critical success factors
   (c) SWOT analysis

(3) *Step three.* How the outcomes arising from self-assessment can be effectively used by consideration of:
   (a) Changes that will be required

   (b) People that will be prepared to champion the changes
   (c) Employees who will support the changes

(4) *Step four.* Implementation of the priority actions by all concerned, remembering to:
   (a) Allocate necessary resources
   (b) Agree plans
   (c) Co-ordinate the input of particular teams/departments or individuals

(5) *Step five.* Review and measurement of effects to consider how effective the implementation of action has been:
   (a) Monitor progress
   (b) Change resource input if required
   (c) Be responsive to external environment and differing circumstances

The EFQM four-phase model for implementation of output of self-assessment

Following research by an EFQM working group, there is a recommended four-phase model for the integration of the output that arises from self-assessment.

(1) *Phase one: Overall.* In this phase, the mission and values of the organisation are reviewed and, subsequent to carrying out self-assessment and use of data from 'traditional' sources, *priority* areas for improvement are identified in conjunction with a general business plan for a predetermined period (i.e. one year). As a result, an integrated plan can be published which will describe what will be done, how, and the forecast of effects on performance indicators.
(2) *Phase two: Identification of improvement plan.* This phase can be considered to be a development of part of phase one, and involves reviewing the sources of data to ensure that the improvement plan is appropriate to the business needs of the organisation.
(3) *Phase three: Review of data.* This phase is dedicated to ensuring that all data being collected from assessment of the 32 sub-criteria are appropriate to the business planning process.
(4) *Phase four: Re-evaluation.* In this phase, the organisation should re-evaluate whether the data that have been collected are having a positive impact upon its ability to create excellence through continuous improvement. As EFQM believes, it

may be necessary for the management of an organisation to realign their statement of its espoused mission and values. Finally, the likelihood of success of organisational improvement will be enhanced by ensuring that every person is made aware of what using the Excellence Model involves and, in particular, how crucial it is that they contribute.

7.6.5 Reviewing the use of the EFQM Excellence Model

As the EFQM Excellence Model becomes increasingly adopted by organisations, it is not surprising that research into its impact has been carried out. Despite what may be interpreted as being a biased opinion, EFQM suggests that there are ample results available from research to 'reveal that Self-Assessment is an efficient and powerful tool to control and monitor organisational development' (EFQM, 1999: p. 52). Moreover, EFQM argues, even though using the Excellence Model will require considerable effort, such commitment 'pays in terms of bottom-line results'. As the case studies presented in the next chapter suggest, the experience of construction organisations that use the EFQM Excellence Model would tend to agree with this statement.

Summary

This chapter has described what any organisation must do in order to attempt to become world class. In particular, the following aspects of world-class status have been explained:

- A definition of what being world class actually involves
- The stages of development that an organisation may go through in the quest to achieve world-class status
- The importance of excellence awards in identifying world-class organisations
- How self-assessment is used in implementing improvement
- The evolution of excellence awards
- The EFQM Excellence award
- How it is possible to utilise the EFQM Excellence award in order to carry out benchmarking to achieve world-class status

CHAPTER EIGHT
MOVING FROM THEORY TO PRACTICE

Objectives

Chapter 1 explained the influence – most especially through the report *Rethinking Construction* (Construction Industry Taskforce, 1998) – that is being brought to bear upon the construction industry to ensure it is more efficient and capable of providing potential clients with products and services that are perceived to be value for money. The book has drawn attention to the fact that certain industries have, by learning from elsewhere, ensured that their customers get the best. In essence, this is what benchmarking involves. Advocates of improvement use as justification for their beliefs, the emergence of Japanese producers of electronic and automotive products who are judged as being pre-eminent in terms of quality and reliability. As these advocates typically explain, the ability of those organisations which produce such goods was achieved by the dedicated use of tools and techniques to ensure that every aspect of what they do is constantly improved. The Japanese use three words which summarise this approach: *Dantotsu*, the constant quest to be regarded as being the best; *Kaizen*, continuous small-step incremental improvement; and *Zenbara*, the constant search for best practice and using it to improve whatever it sells.

As has already been described, the consequence for organisations which use such an approach is, by being able to 'delight' customers (something the Japanese call *Miryokuteki Hinshitsu*), the ability to achieve a perception among customers that they get the best available. This results in increased profits which, in turn, allows further investment in the future development of products or services (something Deming calls the 'chain reaction'). It is important to recognise that another characteristic of such organisations is their treatment of people. Instead of being regarded as simply resources to be used, they are viewed as being absolutely crucial to the long-term development of the organisation. Because of this, there is a commitment to the training and education of employees to allow them to dedicate themselves to ensuring that success through customer satisfaction occurs.

It is precisely these sentiments that advocates of change in the construction industry believe are necessary. CIRIA (Construction Industry Research and Information Association) argues that benchmarking is 'a powerful tool for continuous improvement' (CIRIA, 1998: p. 7). As it suggests, construction can, if it is willing to learn from elsewhere (benchmark), enjoy similar benefits to those that accrue to world-class organisations (CIRIA, 1998). As a consequence, CIRIA recommends that in order to assist in achieving this objective, 'the experience and results of benchmarking exercises in construction should be promoted widely' (CIRIA, 1998: p. 39). As the case studies in this chapter clearly show, there are organisations that have already used the tools and techniques of benchmarking to produce improvement. Subsequent to Chapter 1, the book has described the theories that are essential to the successful implementation of benchmarking:

- Understanding the concepts (Chapter 2)
- TQM (Chapter 3)
- Cultural change (Chapter 4)
- Critical success factors, processes and systems (Chapter 5)
- Measuring customer satisfaction (Chapter 6)
- The importance of using excellence awards to improve (Chapter 7)

This purpose of this chapter is to provide advice and guidance to any manager who, having understood these theories, applies the concept of benchmarking for best practice to produce improvement in his or her organisation. As such, this chapter will consider the following:

- The way to get started in applying the concept of benchmarking
- What are the possible obstacles that should be overcome in attempting to use benchmarking
- A three-phase, fourteen-step model to approaching benchmarking
- A number of case studies

8.1 Where to start from

During the course of researching the subject of quality management, a manager charged with this task told the author the following:

> You know, the funny thing about being involved in quality is that everyone sees you as telling them obvious things; like trying to do things right first time, and making sure that the customer gets what they want. As some tell you, 'Quality is just commonsense'. But you know what, as soon as you go and ask why things have not been done right - and believe me, there will always be things that can be done better - the answer you get is usually, 'I didn't have time', or, 'I did it the best I could'. It's possibly the most frustrating part of this job: getting people to do the obvious.

This story is told as a way of dealing with the potential difficulty of building upon the previous chapters that have described the various theories of benchmarking. In particular, these chapters have explained why benchmarking provides a method which, if it is used appropriately for the situation and context, can be a powerful tool for creating organisational improvement. However, similar to the story told above, the concept, whilst apparently being deceptively simple, will probably be difficult to put into practice. In this respect it is worth remembering that the reason for difficulties may be, like the excuses offered to this quality manager, lack of time or resources. Ultimately, most of what that this book suggests is commonsense. After all, what organisation does not want to achieve its objectives more efficiently and with less cost? More crucially, what manager does not believe that his or her people would - if they were given more time, resources and training - be able and willing to do a better job? These, of course, are obvious. But as the examples of organisations described in preceding chapters indicate, doing things that will ensure obvious benefits such as customer satisfaction and delight, for what seem like compelling reasons at the time, can sometimes be forgotten.

The existence of any 'hurdles' that may militate against carrying out activities aimed at creating improvement must be recognised and dealt with. As Chapter 4 explained, creating a desire for change that will be sustainable, i.e. willingly supported, means that the initiatives must be sufficiently considered, well-planned, and most importantly of all, explained to all who will be involved. Unless people believe there will be benefit both to them and the organisation in general, they will be tempted not to bother. As this chapter will suggest, in order to make the process of benchmarking more likely to succeed, it is necessary to be aware of a number of things. In order to make the transition from theory to practice, this chapter will describe issues that have emerged from the author's research of organisations that have attempted to use benchmarking in order to create improvement. In so doing, they should be regarded as obstacles to be encountered and appropriately dealt with.

8.2 *Obstacles to benchmarking*

Obstacles to benchmarking can be summarised by eleven categories:

(1) Benchmarking only applies to big organisations
(2) The process is going to be too time-consuming
(3) Our customers seem reasonably happy
(4) Counting everything
(5) It must be complex
(6) Industrial tourism
(7) Inadequate preparation prior to benchmarking other organisations
(8) Unrealistic timetable
(9) Team being wrongly composed
(10) Improper emphasis
(11) Limited support by senior managers

Each of these obstacles is discussed in turn in the following subsections.

8.2.1 Benchmarking only applies to big organisations

It is undoubtedly true that this perception is a valid one. As the relevant literature shows, large organisations – particularly multinational companies – tend to be those most frequently used as the case studies. It is understandable why this occurs; the case studies included in this chapter tend to reinforce this view. Whilst it is good that large organisations are willing to lead the way, it is small organisations that make up the major part of the construction industry. The import of this fact, therefore, is that if, as the Egan

Report recommends, the construction industry is to engage in overall improvement, small organisations must be actively engaged in using tools such as benchmarking.

A project carried out by a research team from the University of Central England has assisted small and medium-sized enterprises in the West Midlands to use benchmarking in order to address their ability to achieve customer satisfaction. The experience of those organisations that have collaborated in this project is that merely attempting to think about issues which, hitherto, had largely been taken for granted, is a valuable exercise.

One of the most successful organisations to have used benchmarking was a small quantity surveying firm which employs only five people. This organisation had reached the stage where reorganisation was required to keep up with a workload which, despite consistently improving, was resulting in reduced profitability. As this organisation discovered, they were spending more time on tendering for new work than managing existing contracts. What the managers found after benchmarking against a company which carried out catering, was the need to spend time analysing potential contracts for opportunities to create value-adding. Once the managers of the quantity surveying firm did this, they quickly discovered that it was more profitable to concentrate their effort on a smaller number of customers to which they could provide a much-enhanced service. Moreover, carrying out these contracts was a much more fulfilling experience. As the managing director said:

> It's one of the strange things you find. Previously, we had more work than we could cope with. As a consequence, we were working for clients who expected us to work for nothing. When we would try to claim what we thought was a fair rate for the work, they would argue. Because of what C ... [the catering firm] showed us, we now concentrate our work on those clients where quality, not price, is the main objective. We give them a much better service and they are happy to pay us more. I cannot believe it took us so long to realise what should have been obvious. The trouble is, when you are as busy as we were, you don't have time to think about what you are doing. One of the things I would recommend to any firm is to get out and see a successful outfit carrying out their business. Sure, it will take time, but what you may learn could allow you to make the transition we have achieved.

8.2.2 The process is going to be too time-consuming

The issue of time is one that should be considered at the earliest opportunity. As the cliché goes, 'No pain, no gain'. It is a fact of life that unless those who manage an organisation are willing to invest additional effort into applying benchmarking, it will be likely to fail. Chapter 1 used the experience of Formula One in order to show how those teams which want to be the best are constantly endeavouring to create advantage over the competition. The successful teams or individuals in any context tend to be those who are willing to put in the extra effort which others consider unnecessary. What this frequently means is that there has to be compromise between current performance and effort being dedicated to finding improvement in the future. This is equally true of benchmarking in any organisation. As the next obstacle stresses, even if an organisation believes that what it does currently is adequate, in the future, that may not be enough to satisfy customers.

One of the main reasons why any business initiative tends to fail is lack of time. Some of the firms the author researched during his study of the use of QA (quality assurance) using BS 5750 (superseded by ISO 9000; BSI, 1994) worked on the assumption that all that was required was to make someone responsible for writing a manual of quality procedures. Managers in such firms tended to justify this view by arguing that as long as they could show that the 'plaque was on the wall', they could prove that they had achieved QA. Whilst doing this may have provided short-term comfort, the long-term results were that little, if any, benefit was experienced. Unfortunately, benchmarking, because it is perceived as being an adjunct to quality, suffers from the belief that it will be like QA, i.e. that it can be implemented as an imposed system.

8.2.3 Our customers seem reasonably happy

Complacency is something that affects everyone at various times in their lives; most people would like to believe that they are the best at doing something. The reality, of course, is that whilst most people do things for pleasure, they tend to accept that 'being average' is probably the most they can reasonably achieve. In the modern business world, however, assuming that providing products or services that are average will suffice is, at best, unadventurous, and at worst, highly dangerous. The message that clearly emerges from those organisations which have made the transition to becoming world class is that they faced serious competition, and had they failed to respond to this, their destruction would have been certain (Rank Xerox is probably the best example).

Knowing what customers think about what an organisation provides to them is vital. This is why it is essential to benchmark their satisfaction levels frequently. By so doing, it should be possible to accurately gauge what they believe the organisation does well, and more importantly, what it doesn't do so well. The former should be regarded as 'good practice' (remember that even better practice may exist elsewhere), and the latter are opportunities for improvement. If nothing else, the mere fact that an organisation bothers to find out what its customers think will be interpreted as being considerate.

8.2.4 Counting everything

There is a tendency when commencing benchmarking to believe that by measuring everything, improvement is bound to occur. What is important is that an organisation understands the day-to-day processes and, as a result, what is crucial to the ability to achieve critical success factors (CSFs). This is a bit like the chicken and the egg: which comes first? The simple answer is that if those who are responsible for the benchmarking process sufficiently understand what goes on, they will intuitively know what the organisation seeks to achieve, and what things must be done in order to ensure they occur. For this reason, the use of process mapping and identification of CSFs is vital. Once these have been established, it is less likely that the organisation will engage in 'overmeasuring' by attempting to assess aspects of the organisation which are not crucial to success.

8.2.5 It must be complex

When some people come to the subject, in their desire to understand everything, they may consult texts that provide models and explanations that seem daunting. The fact is though, at the core of the concept of benchmarking is the desire to learn how to do things better. People do this in their everyday lives when they go to be coached to play sports more competitively, to communicate more effectively, or (as the number of television programmes suggest) want to cook, decorate or tend to their gardens better. By mastering the simple things, it is usually possible to increase skill and confidence in the more complex areas. Being in business is no different; organisations can learn to be more effective by comparing their simple day-to-day practices against those who are acknowledged to be either better or the best.

8.2.6 Industrial tourism

This is a term that applies to those who, in their desire to see every world-class organisation possible, conduct the visits in a way that negates the ability to learn anything useful. In particular, because those carrying out benchmarking are not sure of the most important processes that their own organisation carries out, they will not be sufficiently focused in analysing comparable processes in the host organisation. The next obstacle also stresses the need to adequately prepare before visiting other organisations.

8.2.7 Inadequate preparation prior to benchmarking other organisations

Whilst it may be inspiring to walk around a 'state-of-the-art' organisation, unless the visitor has a list of pointed questions that can be used to 'interrogate' whatever processes are being used, the time and effort will be wasted. Therefore, it is essential that there is sufficient effort spent on preparing what will be seen, who to speak to and the questions to be asked. As a result, the data that emerge will prove to be far more valuable to ensuring that benchmarking endeavour is successful.

8.2.8 Unrealistic timetable

There are no rules as to how long benchmarking should take. However, it is unlikely to be a process that can be achieved in a couple of weeks. It must be approached as a task that will involve some (eventually all) employees and at the outset, require, as a minimum, many months. Indeed, as the philosophy of continuous improvement is one that is considered to be never ending, so too should benchmarking be regarded.

8.2.9 Team being wrongly composed

For benchmarking to work, it is essential that those who are involved have the following attributes:

- They will be involved in the implementation of any changes that impact upon day-to-day operations
- They understand the purpose of using benchmarking and, that if it works well, there will be potential benefit to the organisation, individuals and the client
- They have the right skills and have been provided with training that will allow them to be effective in using the tools and techniques

Whilst these points may seem obvious, the belief that people can be forced to use something that they either don't understand, or worse, fear, will cause resentment and undermine the likelihood for success. As this book stressed from the outset (see section 1.2), people are at the heart of benchmarking. The research that the author has carried out into the use of quality management in general shows that misunderstanding the importance of people in attempting to create opportunities for improvement will simply cause more damage than good.

8.2.10 Improper emphasis

If, because of inadequate preparation and training of those people involved in the benchmarking exercise, they are expected to achieve too much or are unclear as to what are the most crucial aspects of processes, the results will be confusion of purpose. The main thing about a sport such as running, for instance, is that the goal is absolutely clear: to be faster than anyone else. Business organisations tend, of course, to be more complicated than simply running races! However, the fact still remains that the corporate graveyard is full of businesses that were not fast enough to cope with change. As a consequence, teams that are given the task of improving particular parts of the business should, even though the aim is overall improvement of the whole organisation, understand the need to be focused on small-scale achievable change.

8.2.11 Limited support by senior managers

Senior management commitment is a frequently quoted maxim with respect to quality management and improvement. However, the research that the author has both carried out and read suggests that sometimes top management forget that whilst their role is to be strategic, they must also ensure that they are seen to lead by example. In any change process, if senior managers are believed to lack interest or are too busy to appreciate the efforts of those at operational level, future initiatives will inevitably suffer from the 'If they don't care, why should I bother syndrome'. At this stage, motivation will have been so badly affected that a tremendous amount of effort will be required to repair the damage done. In the meantime, however, competitor organisations will probably have continued apace their development towards excellence. The consequence, sadly, will be that it becomes even harder to catch up.

8.3 *A three-phase, fourteen-step approach to benchmarking*

This section discusses a three-phase, fourteen-step model for benchmarking. This is illustrated in Table 8.1 and what each of these steps involves is described in more detail below.

8.3.1 Phase one: preparation

Step 1: Obtain management commitment and define CSFs

This hardly needs elaboration at this point. The decision to engage in benchmarking may have resulted from a concern by senior managers that the organisation is falling behind competitors. As a consequence, a new mission may be required to focus all employees on the need to develop parts of the organisation in order to match the level of service and standards provided by competitors.

Regardless of the motivation, the starting point for implementing the tools and techniques associated with benchmarking will emanate from those who are responsible for the stewardship of the organisation at the highest level. As such, it is their responsibility to ensure that the following things are in place:

- Communication of the need to improve and how
- The resources that will be necessary to support everyone's efforts
- Finance adequate to achieve objectives
- Time available to devote to the use of benchmarking
- Training and development for every person who believes they need assistance or new skills
- A definition of the CSFs for the organisation – these will provide the focus for achieving corporate objectives

Table 8.1 The three-phase, fourteen-step model

Phase (personnel responsible shown in italics)	Steps
Preparation	
Management	1. Obtain management commitment and define CSFs
Process 'operators' Management	2. Understand your own processes 3. Identify strengths and weaknesses in processes
Management	4. Select processes to be benchmarked 5. Create benchmarking teams
Teams	6. Consider which are the best in the organisation's processes to be improved 7. Select an organisation against which to benchmark
Execution	
Teams with partners	8. Establish benchmarking agreement with 'host' organisation 9. Carry out benchmarking in order to gain information data and advice on process improvement
Teams augmented as required	10. Analyse information and data collected in order to assess gap between existing processes and best practice 11. Plan actions to close gaps and improve processes 12. Implement agreed changes to create improvement
Post execution	
Management	13. Constantly monitor and measure the impact of changes implemented 14. Set new benchmarks and continue the process by considering other processes to be improved

Step 2: Understand your own processes

This was dealt with in Chapter 5. Unless those who will be involved in improvement are sufficiently understanding of what they do, how will it be possible to learn from others the way to institute changes in day-to-day processes that will lead to real and sustainable improvement? As explained in Chapter 5 mapping processes is a valuable exercise for learning what goes on and why. It should be anticipated that what people frequently *think* goes on is different to what actually *does* go on. Mapping processes – an exercise which, because it is diagrammatic, will be easy to understand - will show if any such differences exist.

Step 3: Identify strengths and weaknesses in processes

Learning the strengths and weaknesses of day-to-day processes is essential in benchmarking. The strengths should be celebrated as existing good practice (even better may exist elsewhere), and the weaknesses as 'opportunities for improvement'. Using benchmarking will assist in ensuring that the former are as good as the best, and that the latter are also improved to be equally so.

Step 4: Select processes to be benchmarked

Following on from step 3, it is important that processes are selected which, it is believed, can be improved by benchmarking. It is likely that they will be selected upon the basis of:

- Being crucial to the development of the organisation's ability to provide excellent service and standards to customers
- Being those that are in most need of radical change and improvement

Step 5: Create benchmarking teams

The importance of creating the right team was described in point 9 of the obstacles to benchmarking (section 8.2.9). In addition, section 4.4.2 of Chapter 4 deals with the importance of developing teamwork as the basis for creating cultural change. As such, this is a part of the process of benchmarking that requires management in order to ensure that those who are involved have the skill, confidence and motivation so as to devote their fullest efforts to the objective of improvement.

Step 6: Consider who are the best organisations in the processes to be improved

At this point an organisation will be considering the sort of organisations (or different parts thereof, if the benchmarking is internal) to visit where processes considered to be exemplary exist. This is not easy. There will be issues of access and confidentiality to be considered (see step 8). More difficult for any organisation, is knowing which organisations might be potentially useful for this process. Clearly, for internal or competitive benchmarking, departments or competitors may be used (in the case of the latter, unless they operate in a different geographical region, there will be sensitivity). However, in the case of functional or generic benchmarking where organisations that are considered to be the best are used, an obvious place to search will be the list of recent winners of a quality award or a regional variant of the EFQM Excellence Model. Those organisations that win these awards are usually very happy to allow others to visit their organisations for the purpose of benchmarking their processes.

Step 7: Select an organisation against which to benchmark

Following on from the decision-making of step 6, it will be necessary to make contact with an organisation that will allow benchmarking against their processes. Given the amount of time and effort that each party will put into the process, it is important to consider as fully as possible every aspect of how this relationship will operate. The more successful the relationship, the more likely it is that each will get something useful from the exercise.

In addition, the following things should be considered:

- Time required (select an organisation that is close, so as not to involve considerable travel)
- Will there be the potential for benefit to both sides? (Excellent organisations are always aware of the possibility of learning from others)

8.3.2 Phase two: execution

Step 8: Establish benchmarking agreement with 'host' organisation

It will be extremely valuable to consider all the likely issues that will be involved in ensuring that the benchmarking can be carried out effectively and efficiently. There will have to be some allowance for disruption to each of the organisations when visits occur. As a consequence, a protocol for the benchmarking arrangements should be agreed in order to avoid potential conflict occurring later; this would include:

- Disclosure of information
- Use of data/copyright material
- Access to certain departments and people

Step 9: Carry out benchmarking in order to gain information, data and advice on process improvement

The need to collect data that will allow understanding of the host organisation's processes is, of course, the key objective of benchmarking. In order to achieve this, those who are involved should constantly endeavour to discover the answers to the following questions:

- What factors make the processes of the host organisations so good?
- How do they utilise the factors of production?
- Are people organised in such a way as to make the host organisation radical?
- What knowledge can be transferred successfully to improve operations and processes?
- What has been learned that may assist not only the day-to-day processes being specifically considered, but could also be more widely applied?

Step 10: Analyse information and data collected in order to assess gap between existing processes and best practice

Having carried out the visit to the host organisation and as point 6 of the obstacles to benchmarking suggested (industrial tourism), how can the information and data be used in such a way as to increase the chances that any changes initiated will result in improvement? By identifying gaps between the existing processes and those of the host organisation and, knowing why the latter is so good, it should be possible to consider how to deal with step 11 – planning how to do things differently.

Step 11: Plan actions to close gaps and improve processes

As with all other steps, careful consideration will be required; there is no point making rapid changes to processes if all that happens is the opposite to improvement. Unfortunately, there are no guarantees in anything in business (even in benchmarking). However, as has been continually stressed in this book, constant measurement will allow immediate assessment of the impact of actions (see step 13). In planning the action, it is essential that there is a consensus of agreement from all those directly involved, most especially those who will be involved in the implementation of changes to day-to-day processes. The support of people at every level (particularly operational) is essential in creating a culture in which changes can be implemented to create organisational improvement.

Step 12: Implement agreed changes to create improvement

At this stage, implementation of changes should occur automatically; there should not be any issue over lack of time or resources (if that is the case, the whole effort will have been a waste of time). Much will depend upon what is required to implement the change. There will be no limit to the potential changes that may emerge, for example:

- Additional staff to deal with problem areas
- Training or development of existing staff
- New equipment
- Alterations to established practices
- Alternative systems for controlling processes
- Reorganisation of departments
- Faster/more robust methods of communication
- Enhanced reporting of customer problems

What is important, of course, is to ensure that the result of change is improvement. If there are any negative effects because of the change, immediate action may be necessary to ascertain whether they are 'temporary teething problems' and will soon be rectified. However, readers should remember their own experiences of problems with a product or service that have been explained as 'temporary teething problems'; customers will not be terribly interested in the inability of the organisation to get things right first time. A dissatisfied customer is less likely to come back and therefore, no matter how good an organisation's product or process is in the future, it may be impossible to attract their custom again.

8.3.3 Phase three: post execution

Step 13: Constantly monitor and measure the impact of changes implemented

Figure 8.1 summarises what should be the impact of introducing changes to a process. Normally there will be a fairly rapid

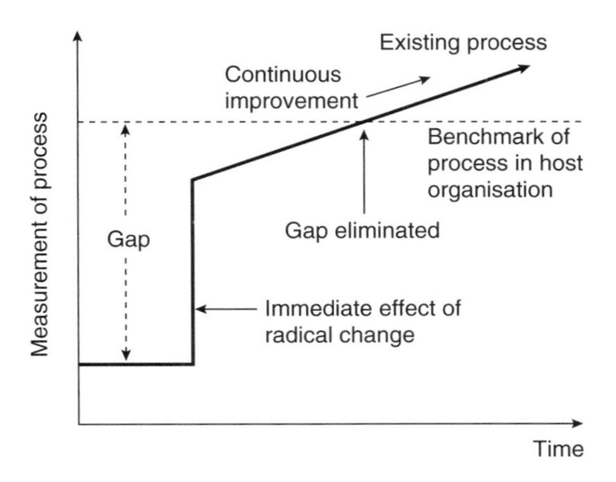

Fig. 8.1 Gap reduction following benchmarking of existing process.

improvement. In the long-term, however, this should be followed by efforts to continuously improve.

Step 14: Set new benchmarks and continue the process by considering other processes to be improved

The need to engage in improvement should be seen as something that is never-ending – there is no completion. As previous chapters have described, excellence is something that cannot be taken for granted; business, like sport, is far more competitive than was the case in the past. The world-class organisations of today are constantly thinking of how to be even better tomorrow. They know that if they do not do this, there are others who will try to beat them in terms of reliability, value or cost; hence the cliché that tends to be used in considering excellence, 'Get better or get beaten'. Therefore, unless an organisation is striving to be better in the future, its chances of being the number one choice will be severely limited.

8.4 Other aspects to consider

There are additional things that can be done in order to assist in applying the concepts that have been described above. One of the most important is not to feel that you are alone. Points of contact are discussed in the following subsections:

- Benchmarking clubs
- How to start the process of utilising the EFQM Excellence Award

- Useful addresses
- Further reading

8.4.1 Benchmarking clubs

Pickrell *et al.*, in their publication *Measuring Up: a practical guide to benchmarking in construction*, assert that benchmarking clubs offer an 'easier route' to using this technique, 'particularly for smaller companies' (Pickrell *et al.*, 1997: p. 60). As they explain, such clubs allow members to share problems that affect a certain trade or profession. As such, if an organisation wishes to learn about benchmarking, these clubs allow a method by which, provided that the organisation is willing to share expertise and practices, it can improve processes. Benchmarking clubs currently exist around the country for organisations that wish to benchmark against other organisations, regardless of the sector in which they operate and, specifically, for those that wish to exchange information about the construction industry. The addresses of benchmarking clubs are given in section 8.4.3 and further information can be obtained from the Construction Best Practice Programme (whose address is also included).

The author is a member of a group largely consisting of contractors operating in the Midlands known as the Midlands Construction Forum. This group, which has been in existence for over five years, was originally formed to allow quality managers to discuss matters specifically relating to QA using BS 5750/ISO 9000 (BSI, 1994). However, as quickly became evident in the first couple of meetings, some organisations had already considered alternative approaches to ensuring that they were capable of pursuing excellence in terms of customer satisfaction and quality of the finished product. Those organisations that regularly attend meetings of this Forum are able to constantly learn from the experiences of others. As this Forum has developed – entirely unfunded by any outside organisation it should be stressed – it has been able to raise its profile as a body concerned with supporting national initiatives to improve the quality of British construction. As a consequence, its views are now routinely sought by representatives of research groups and Government-sponsored bodies such as the DETR, CIC (Construction Industry Council) and BRE (Building Research Establishment).

One of the aspects of quality management that has concerned the Midlands Construction Forum in recent years has been how to engage in benchmarking. In order to assist members in benchmarking, various exercises and discussions have been organised. Most particularly, a number of members are actively pursuing implementation of the EFQM Excellence Model in their organisations. As Chapter 7 described, this is widely acknowledged to be the method by which a client can ascertain which potential supplier organisations are capable of meeting their expectations. As a consequence, it is recommended that every organisation should consider how it might use this model.

8.4.2 How to start the process of utilising the EFQM Excellence Model

Whilst the EFQM Excellence Model, as described in Chapter 7, may appear daunting at the outset, it is important to remember that its component parts are based upon the logic that any good organisation should aspire to. These components are no less applicable to construction than any other sector of industry. Crucially, in the author's research experience and as some of the practitioners who provide case studies for this book are finding, clients are increasingly making the implementation of this model a prerequisite to being allowed to tender for potential contracts. Therefore, knowing how to apply the principles of the EFQM Excellence Model to your organisation will become more and more essential in being able to maintain a competitive position.

In order to get started in using the EFQM Excellence Model, it is necessary to consider *where* to start. In order to do this, it is recommended that those embarking on this process should contact either the British Quality Foundation (BQF) or EFQM (see section 8.4.3). The BQF will be able to provide a great deal of information as to where you might commence the process of using the EFQM Excellence Model. In particular, there exists what are known as the BQF Rapidscore and Teamscore. As the BQF explains in a leaflet to propagate these tools:

> They are powerful diagnostic tools that help you understand your organisation's capabilities and its opportunities for improvement. Both graphically illustrate maturity profiles against the disciplines of the Model.

The reader's attention is also drawn to the easy-to-use guidance manuals *The Excellence Routefinder* (Hakes *et al.*, 1999) and *The Business Driver* (Bristol Quality Centre, 1999). (See further reading section 8.4.4 for full references.) These provide advice on how any organisation should consider the easiest way to start the process of self-assessment in order to align with the EFQM Excellence Model.

As has been consistently stressed throughout this book, the best way to learn how to do something is to find out how others do it; that is, to benchmark. This might be through the benchmarking clubs already recommended. However, there is an organisation which organises a series of workshops and seminars which enable any person or organisation to listen to the experiences of others using tools and techniques associated with benchmarking (use of the EFQM Excellence Model is regularly featured). This organisation is the Construction Productivity Network, and details of how to contact it can be found in the next section.

Finally, there can be no better way to learn about how the EFQM Excellence Model is used than to do so by becoming an assessor. Every region of the UK operates this model and requires individuals who are prepared to be assessors. In order to become an assessor, it is necessary to go on a training course provided for the relevant region. These are provided at a fairly reasonable cost. The benefit that will be received is an intimate knowledge of what the model contains and how it should be scored. The best way of discovering the regional training bodies is to contact the British Quality Foundation, details of which are provided in the next section.

8.4.3 Useful addresses

British Quality Foundation (BQF)
32–44 Great Peter Street
London
SW1P 2QX
Tel: 020 7654 5000
Fax: 020 7654 5001
Website: www.quality-foundation.co.uk
e-mail: mail@quality-foundation.co.uk

CALIBRE
Building Research Establishment
Bucknalls Lane
Garston
Watford
Herts
WD2 7JR

Tel: 01923 664257
Fax: 01923 664398
e-mail: chrysostomouv@bre.co.uk

Construction Best Practice Programme (CBPP)
PO Box 147
Watford
Herts
WD2 7RE
Telephone Helpdesk: 0845 605 5556
Fax: 01923 664290
Website: www.cbpp.org.uk
e-mail: pmu@cbpp.org.uk

Construction Productivity Network
6 Storey's Gate
Westminster
London
SW1P 3AU
Telephone Helpdesk: 020 7222 8891
Fax: 020 7222 1708
Website: www.ciria.org.uk
e-mail: gareth.thomas@ciria.org.uk

Construction Round Table
c/o Building Research Establishment
PO Box 492
Bucknalls Lane
Garston
Watford
Herts
WD2 7JR
Tel: 01923 664378
Fax: 01923 664379
Website: www.crt.org.uk
e-mail: treadawayk@bre.co.uk

Design Build Foundation
Unit 25
University of Reading
London Road
Reading
RG1 5AQ

Tel: 0118 931 8190
Fax: 0118 975 0404
Website: dbf-web.co.uk

CII/ECI Benchmarking Initiative
European Construction Industry Benchmarking Institute
Arnold Hall Building
Loughborough University
Loughborough
Leicestershire LE11 3TU
Tel: 01509 223526
Fax: 01509 260118
Website: eci-online.org
e-mail: eci@lboro.ac.uk

European Foundation for Quality Management (EFQM)
Brussels Representative Office
Avenue des Pléiades 15
1200 Brussels
Belgium
Tel: +32 (2) 775 35 11
Fax: +32 (2) 775 35 35
Website: www.efqm.org
e-mail: info@efqm.org

Government Construction Clients Panel Benchmarking Group
HM Treasury
Allington Towers
19 Allington Street
London
SW1E 5EB
Tel: 020 7270 1624
Fax: 020 7270 1639
Website: hm-treasury.gov.uk/pub/html/gccp/
e-mail: charles.botsford@hm-treasury.gov.uk

Major Contractors Group Benchmarking Club
Construction House
56–64 Leonard Street
London
EC2A 4JX
Tel: 020 7608 5140
Fax: 020 7608 5141
e-mail: mcg@mcg.org.uk

National Contractors Federation Benchmarking Club
Construction House
56–64 Leonard Street
London
EC2A 4JX
Tel: 020 7608 5144
Fax: 020 7608 5141

Workplace Best Practice Group
AWA Best Practice Benchmarking Limited
148 Leadenhall Street
London
EC3V 4QT
Tel: 020 7743 7112
Fax: 020 7743 7111
Website: advanced-workplace.com

8.4.4 Further reading

The following texts are, in addition to the references contained at the end of this book, intended to provide recommended reading to those who want to discover more about the principles that have been discussed throughout this book. It is stressed that because of the complexity of the subject matter, no single text will provide everything that readers need to know about every aspect of benchmarking for best practice. However, the more that readers are able to read about the subject, the more they are likely to learn – a key component of using the tools and techniques of benchmarking – and as a consequence, increase their chances of success.

In order to assist in doing this, the texts have been categorised under the following headings:

- General reading (including benchmarking and relationship marketing)
- TQM and managing cultural change
- Customer satisfaction and world class
- EFQM Excellence Model Award

General reading (including benchmarking and relationship marketing)

McNair, C.J. & Leibfried, K.H.J. (1992) *Benchmarking, a Tool for Continuous Improvement*. John Wiley and Sons, New York.

Payne, A., Christopher, M., Clark, M. & Peck, H. (1998) *Relationship Marketing for Competitive Advantage*. Butterworth-Heinemann, Oxford.

Rolstadas, A. (1995) *Benchmarking – theory and practice*. Chapman and Hall on behalf of the International Federation for Information Processing, London.

Zairi, M. (1998) *Effective Management of Benchmarking Projects – Practical Guidelines and Examples of Best Practice*. Butterworth-Heinemann, Oxford.

TQM, teamwork and managing cultural change

Atkinson, P.E. (1990) *Creating Culture Change: The Key to Successful Total Quality Management*. IFS Limited, Bedford.

Baden Hellard, R. (1993) *Total Quality in Construction Projects, Achieving Profitability with Customer Satisfaction*. Thomas Telford, London.

Baden Hellard, R. (1995) *Project Partnering: Principles and Practice*. Thomas Telford, London.

Brown, M.G., Hitchcock, D.E. & Willard, M.L. (1994) *Why TQM Fails, and What to Do about It*. Irwin Professional Publishing, New York.

Carnall, C. (1995) *Managing Change in Organizations*. Prentice Hall, Hemel Hempstead, Herts.

Fombrom, C.J. (1992) *Leading Corporate Change, How the World's Foremost Companies are Launching Revolutionary Change*. McGraw Hill, New York.

Lamming, R. (1993) *Beyond Partnership, Strategies for Innovation and Lean Supply*. Prentice Hall, Hemel Hempstead.

McCalman, J. & Paton, R.A. (1992) *Change Management, a Guide to Effective Implementation*. Paul Chapman Publishing Limited, London.

Murata, K. & Harrison, A. (1991) *How to Make Japanese Management Methods Work in the West*. Gower Publishing Limited, Aldershot.

Oakland, J.S. & Porter, L. (1994) *Cases in Total Quality Management*. Butterworth-Heinemann, Oxford.

Pettigrew, A. & Whipp, R. (1993) *Managing Change for Competitive Success*. Blackwell Publishers, Oxford.

Taylor, B. (1994) *Successful Change Strategies – Chief Executives in Action*. Director Books, Hemel Hempstead.

Wille, E. (1992) *Quality: Achieving Excellence*. Century Business, London.

Ward, M. (1994) *Why Your Corporate Change Isn't Working . . . and What To Do about It*. Gower Publishing Limited, Aldershot.

Customer satisfaction and world class

Dimancescu, D. & Dwenger, K. (1996) *World-class New Product Development, Benchmarking Best Practice of Agile Manufactures*. American Management Association, New York.

Morton, C. (1994) *Becoming World Class*. MacMillan Press Limited, Basingstoke.

Presscott, B.D. (1995) *Creating a World Class Organisation.* Kogan Page, London.
Pümpin, C. (1993) *How World Class Companies Became World Class.* Gower Publishing Limited, Aldershot.
Walker, D. (1990) *Customer First, a Strategy for Quality Service.* Gower Publishing Limited, Aldershot.

EFQM Excellence Model Award

Bristol Quality Centre (1999) *The Business Driver, a Performance Reviev Programme.* Bristol Quality Centre, Bristol.
Hakes, C., Bratt, S. Norris, G., Wildman, T., Parry, M. & Gallacher, H. (1999) *The Excellence Routefinder – Helping You to Create an Agile Organisation Ready to Face the Challenges of the Future.* Bristol Quality Centre, Bristol.
Zinc, K.J. (1997) *Successful TQM – Inside Stories from European Quality Award Winners.* Gower Publishing, Aldershot.

8.5 *Case studies of benchmarking in construction organisations*

The case studies provided in this section are written by practitioners who, using various tools and techniques associated with benchmarking, describe their own experiences. These studies are intended to provide encouragement and advice on how readers might attempt to use the methods of benchmarking for best practice in their organisation. In total, there are nine case studies provided.

Case studies one and two are written respectively by Rachel Timms and Keith McGory of AMEC Capital Projects Limited and describe the use of key performance indicators (Rachel, it should be noted, is describing experiences she gained prior to joining AMEC).

In case study three, Nicola Thompson describes how Miller Civil Engineering have responded to the Egan Report by instituting benchmarking through their involvement with benchmarking clubs and the use of tools to produce continuous improvement.

The fourth case study is written by Michael Collini of Hilton International Hotels and explains how he is involved in the constant search for best practice by benchmarking the construction of their leisure facilities.

Case studies five, six and seven are written by Mark Evans (Barhale Construction), Hamish Robertson (Morrison's) and Martin Brown (John Mowlem) respectively. These writers describe how their companies have adopted the EFQM Excellence Model, and how, by using it, they are able to create improvement that will allow them to be compared to the best organisations in the world. Case study eight is written by Lisa Harris and Chris Sykes of AMEY Supply Chain Services Limited who explain how benchmarking tools and techniques should be applied by construction to ensure that consistent client satisfaction occurs.

The final case study is a description of how benchmarking of engineering projects is carried out by the Construction Industry Institute/European Construction Benchmarking Industry Initiative.

8.5.1 Case study one: The use of key performance indicators on contracts for a utilities sector client

Rachel Timms, Quality Co-ordinator, AMEC Capital Projects Limited

Overview

As a business studies graduate, I have been involved in the implementation of various aspects of business in the construction industry. What I shall describe is my experience of having been involved in the use of key performance indicators (KPIs) in the utilities sector during my experience at another major contracting company.

The client

It helps to have an enlightened client. The client alluded to here was from the telecommunications industry and was certainly one of the most forward-thinking, yet demanding, clients I have ever come across. As a consequence of the Latham and Egan Reports, this particular blue-chip client was already forging ahead with its own approach to prime contracting, benchmarking and partnering through its Vendor Rating Programme.

The client needed to improve its infrastructure provision in Britain to meet with a massive rise in personal computing, digital networks, mobile phones and interactive communications. Therefore, it needed contractors to deal with maintenance, rerouteing, and new infrastructure provision for fibre-optic cables and older network systems to meet with such demand.

The contractors were chosen on ability, innovation, commitment and price. The client didn't just want the lowest price but perceived added value of long-term arrangements which could ultimately bring economies of scale to all parties involved in the supply chain, including subcontractors. This approach from the client was part of a larger strategy to bring about preferred supplier status in the future. Benchmarking provided the means to evaluate the client's own supplier base for provision of utility services.

The contractor

As one of the largest contractors in the construction industry, diversification into new markets, including utilities, was a necessary strategic development in order that this global player could strengthen its customer base through diversification. The aim was to provide total supply chain solutions for clients. The initial problem was direct competition with many small to medium-size contractors; i.e. those who have traditionally had a stronghold in this sector. As a 'big fish in a small pond', it was important to the company to gain a vital market share in the sector. Therefore, a partnering approach with the client on a fixed-term contract was negotiated – something that created a less adversarial and traditional client–contractor relationship. As we experienced, a more open and honest approach to business was envisaged.

The contract was negotiated over a five-year fixed term for provision of infrastructure maintenance and development services to this major blue-chip client. This represented initially a commitment of at least £55 million per annum to the contract which was based on three different geographical zones – Southern Home Counties, Northern Home Counties and East Midlands. As the client was already evaluating our performance through its own vendor rating system, it was decided that in order to keep ahead of our competitors, the contract should be carried out using continuous improvement through monitoring of its own performance. Additionally, it should be stated that we anticipated that the enlightened and demanding client would see KPIs as being an essential prerequisite to tendering for contracts.

Under guidance from the client and quality manager, the contract team decided that they wanted to introduce a performance system to monitor key criteria to demonstrate, internally to its own management, and externally to the client, how effective they were. Although obviously the contract had its own departmental systems such as cost management or planning, there was not a tailor-made system for reporting on areas of the contract regarding comparative performance measurement.

In order to achieve this, the quality manager needed to secure senior management commitment to the project by conducting a briefing session in which to communicate how effective benchmarking could be to the contract. This session involved not only senior contract management, but also the operations director for the region. The briefing session proved favourable and, as a consequence, it was decided to introduce a KPI system. The operations director provided commitment through leadership as a champion and through a written statement of intent that was communicated to all contract employees to encourage involvement and openness to the endeavour.

Within the contract it was received favourably. However, the usual questions remained. Would this deliver reductions in waste, rework and defects? Would it lead to more effective management of the processes involved? Would it help with objective-setting and targets? There was, we realised, an underlying feeling that acceptance had more to do with internal competitive issues than a real need to improve.

The employees

The most important step in introducing KPIs was to communicate to all employees on the contract (including subcontractor's representatives) how the initiative would involve them. The quality manager did this by touring each zone and facilitating focus groups to discuss and choose key criteria to benchmark against. These focus groups consisted of a cross-functional team in each area. Each team was made up of the following:

- A zone project manager
- Planning manager
- Operations manager
- Site supervisor
- Operatives
- Administration assistant
- Accounts clerk
- Subcontractor representatives

The quality manager facilitated these sessions to elicit from the team what processes should be benchmarked and which key criteria were important for benchmarking purposes.

Initially, the contract employees were very wary of the idea of benchmarking processes and were very suspicious of the motives for using them. They tended to ask questions such as: was this just another policing mechanism to pinpoint bad performance; would poor performers be identified? While most people were seen to be enthusiastic, there was a feeling that behind all the hype nothing of substance would emerge. There were various misunderstandings about its intentions, particularly from those suspicious that monitoring performance could bring about job losses, and that it was to do with managing simply by numbers. So, from the start, it was not only an uphill struggle to gain commitment, but a cultural shift in order to bring about the change that was required. In order to do this, senior management commitment was essential from the outset so as to generate enthusiasm from the employees and to show that their efforts towards continuous improvement on this contract were valued and would make a difference.

How the criteria for KPIs were decided upon

The criteria for the KPIs to be used were decided upon by focus groups and senior management. Once discussion and communication started, the pace of events moved fast and many sceptics were convinced enough to become committed enthusiasts. How this was done involved the teams being asked a series of fundamental questions using a brainstorming approach. We had to know where we were currently at in terms of reporting systems and what kind of information we needed in order to identify gaps in the current reporting system. One good thing about human nature is that we tend to be able to cope well with the negative aspects of our jobs. By working through these and other issues, we were then able to group together common process areas important to the job in hand.

Performance definitions were determined for each process area; these could be called sub-processes which were given weightings by groups so that their importance and relevance became clear. Having produced new ideas we ranked them in order by consensus. These results were then taken away and compared to come up with an overall KPI system for the contracts. In designing the KPI system, the quality manager facilitated the groups and used further brainstorming, nominal group technique, fishbone diagrams and contingency approaches to aid the development of the system. As management tools, these proved useful to formalising and bringing structure to the facilitation exercise.

As a result, the contract came up with 11 KPIs for the overall system to be run in three operational areas. These KPIs were as follows:

(1) *Percentage quality checks* – the percentage passes of internal quality control checks by supervisors in that month
(2) *Percentage core sampling* – the percentage passes of laboratory core sampling checks by independent external inspector per month. Recognising the time span taken to test, these figures were acknowledged to be the result two months previous
(3) *Number of local authority defects* – the number of local authority defects issued for that month for works undertaken
(4) *Rework value (£)* – the monetary value of works having to be redone in that month quantified through the commercial department
(5) *Percentage performance* – the percentage performance of jobs complete in that month based on the schedule budget versus actual completion
(6) *Number of complaints* – the number of complaints that month from sources such as the general public, local council and subcontractors
(7) *Number of accidents* – number of accidents that month as Health and Safety Executive (HSE) reportable and HSE non-reportable definitions
(8) *Number of training man-days* – number of training man-days that month
(9) *Percentage employee churn rate* – the percentage number of starters and leavers that month to indicate employee retention
(10) *Percentage shortfall in roll-up value* – the percentage number of payments received versus payments outstanding to indicate any shortfalls in payment for that month
(11) *Number of utility damages* – the number of reported utility strikes that month

Once the KPIs were designed, a formalised system had to be set up. Senior management involvement was once again important to brief employees on the findings of the focus groups and KPIs to be used for the contract. In the event, all the KPIs were accepted as relevant and useful to the business and a commitment was shown by all to the implementation, development and reporting of them. Each region volunteered a manager to act as focal point for the collation of the statistics and reporting on a monthly basis to head office. A central co-ordinator was then appointed to administrate and co-ordinate the system under the direction of the quality manager. For this purpose, I was chosen, and hence my insight into how to run a KPI system.

After responsibilities were set up for running the system, the first task of each KPI system manager was to brief those regional employees affected by the new system. It was important at this point to gain employees' trust in a new system that was sometimes regarded with scepticism and suspicion. It is important to stress that the client was briefed because they formed an important output of the system. The operations director and a senior manager met with client representatives to introduce them to what we were doing (a public relations exercise).

The system

The next problem was to decide upon the best format for reporting purposes. We now knew which KPIs were to be used, we had senior management commitment, we had employees suitably briefed and the client happy with the advances we were making, but the system was not operating. This required time and needed considerable input from the centralised IT function in order to use their expertise. It was decided that because this was an experimental system, it would be prepared and designed in a spreadsheet format, and therefore would require minimal maintenance (most office personnel have knowledge of using spreadsheets).

Once the spreadsheet had been designed, the lines of communication and reporting formats were set up. Three regional administration points of contact were set up to communicate the data needed on a monthly basis. Deadlines were set for communication each month in order that the centralised co-ordinator could input the data. The output from this was then put forward in time for the project monthly management meeting for review and action by senior management and the operations director. Each region then got a breakdown of results for their own area to take back to their own regional meetings for action. It was also decided that there would be a six-monthly feedback report to monitor performance and compare results and an overall yearly report to the board. To launch the system, an internal exercise was started to communicate the introduction of the KPI system. This consisted of memos, internal in-house journal articles, posters, briefings and feedback sessions.

The results

Like most experimental systems, there were hiccups and problems along the way. The first was a lack of commitment; this had been envisaged by senior management from the outset, but was now a reality. It was necessary to continually stress the need to be involved in collecting and reporting the data. Another problem of running the system successfully was the underestimation and planning for staff holidays. It was found that in the first summer of its implementation we had not prepared ourselves for the fact that people are allowed holidays; this can create problems, if the only person who possesses specific knowledge or information is absent. Obviously, this was a steep learning curve and was a clear example of an improvement opportunity.

The results in the first three months were fairly dismal, with each region only managing to report half of the KPIs. This, we realised, was due to time constraints or lack of knowledge. So, obviously, the commitment people gave was 'lip service'. In order to address this situation, the operations director was briefed about this apparent lack of commitment to divulge information, while at the same time workshops were held in all regions explaining where and how to find the necessary information and how to report it. The combination of 'carrot and stick' worked well, and the next three months saw the KPI system operating better, with data and results being submitted on time.

The first six months of the system was the period in which most of the mistakes were made and the problems arose. In terms of the first six months' results, these were negligible, as a comparative analysis could not be used due to not having enough data, misunderstandings as to how to report them, and incorrect data being supplied. It was around this time that targets to indices were introduced.

Once the communication and technical problems were ironed out, the KPI system took on a life of its own. It became a disciplined, realistic performance monitoring system. The data now came in regularly; there was less suspicion of it from employees; and as it was seen as a tracking system, project managers saw it not only as a monitoring system, but a way to get issues visible and resolved through a formalised reporting system. Therefore, piecemeal continuous improvement was happening. As a result, improvements to certain processes started to occur. However, in the first year, whilst results were met with warm enthusiasm from top management, scant action was taken. It took extra effort for those at the top to recognise that on a regional basis, the information could be used to compare KPI trends and forecast future performance based on the results. By doing this, attention was drawn towards the indices where a marked decline was taking place and root-cause analysis employed to find out the reasons why particular KPIs on certain contracts underperformed. Correspondingly, for some KPIs, it was possible to detect a marked decline in reworks which exactly matched a decline in local authority defects and utility damages. The ensuing remedy resulted in an increase in financial returns for using the system; something that makes people pay attention to future improvement.

After 18 months, the results reporting system worked very well. Senior management were taking decisions based on the sound and reliable measures reported to them through the KPIs. Processes were being evaluated to consider improvements to the carrying out of operations. Indeed, other areas of the organisation were contacting us to inquire about the use of KPIs. Costs on contracts using KPIs were decreasing. All involved - client, subcontractor and contractor - believed in the system and were willing to publically support its continued use.

Summary

In summary, the road to improvement is never easy. It takes a lot of pre-planning, time, thought and effort. KPIs are just one of a number of ways to improve processes at work in your business. However, when one looks at stark profit margins in construction, any improvement must be worthwhile. The experience gained in this example of using KPIs has demonstrated that they truly assist any business to improve the standards of work delivered to a client. As such, KPIs are an essential tool in carrying out benchmarking.

8.5.2 Case study two: Using project key performance indicators as a tool for benchmarking and best practice in AMEC

Keith McGory, Project Manager, AMEC Capital Projects Limited

My experience

I am a project manager for AMEC Capital Projects Ltd with specific responsibility for the performance of all engineering projects and alliances in the pharmaceutical, fine chemical and industrial manufacturing market sectors. Prior to this, I was a senior project manager for 18 years in project management with Costain Construction, latterly as general manager of their project design office at Stanlow. As a founder team member of the Active Initiative Supply Chain Best Practice Group, I have now been involved in the identification and implementation of best practice and continuous improvement for the last five years.

Introduction

At AMEC, we are continually looking for improvement in the performance of our project and alliance teams. Often the sole criterion is the bottom-line profit and whilst this is the factor that we are ultimately judged upon, there are many factors that contribute to the success or failure of a project/alliance. Regularly, project performance is not analysed unless it loses money or has been deemed a failure. Unfortunately, this does not occur until after project completion. This makes comparisons between project team performance difficult to analyse and measure. It is vital that activities that performed well on one project are incorporated immediately into others. Conversely, activities that were done badly should not be repeated. Identification of these activities can often be achieved by undertaking the project critique, but this is usually carried out at the conclusion of the project when everyone is busy on other work.

Better ways of measuring project performance had to be developed resulting in the production of Project Key Performance Indicators (PKPIs). As part of my involvement in the Active Initiative, I utilised the work undertaken to develop a five-stage project assessment in order to benchmark our project performance and compare this to industry norms. However, this is only part of the picture and within the engineering group there are other factors that contribute to the profit performance. A different set of performance measures were needed in order to address the following issues:

- Areas that need improvement
- Benchmarking our performance against other companies

The performance indicators selected are as follows:

(1) Productivity – based on project man-hour productivity
(2) Performance – based on project man-hour performance
(3) Capital growth – based on project cost over/under-runs
(4) Quality – based on non-compliances per audit
(5) Tender costs – based on tender man-hours per £1000 of contracts won
(6) Safety – based on accident frequency rates

(7) Resignations – based on number of staff resignations
(8) Qualifications – based on staff qualifications
(9) Financial – based on profit variance from budget
(10) Programme – based on project end-date over/under-run

All of the above information was readily available and a sponsor who was both guardian and collator was appointed to produce the report on a quarterly basis. This person had both access to and understanding of the information.

Individual KPI trends

Each of the KPIs selected is now discussed in more detail.

(1) *Productivity.* At a given cut-off point every quarter, add the total man-hours earned to date for all "live" projects and divide by the total (equivalent) man-hours expended to date for the same projects. This gave the overall productivity for all engineering, design and procurement activities. The source of the data was the project progress reports produced by the project planner. The eventual target will be 100% and, subsequently, to reduce the man-hour norms whilst still achieving 100% productivity.

(2) *Performance.* At a given cut-off point every quarter, add the total anticipated final man-hours for all live projects, divide by the total current man-hour budgets for the same projects and express as a percentage. This gave the overall performance in terms of man-hour expenditure compared to the original budgets and is influenced by productivity, programme and change control. The source of the data was the project progress reports produced by the project planner. The initial target was to drive the performance percentage below 100% and thereafter continue towards lower percentages (95%, 90%, 70%, etc.).

(3) *Capital growth.* At a given cut-off point every quarter, add the total anticipated cost over- or under-runs for all purchase orders and subcontracts on all live projects, divide by the current purchase order and subcontract budgets for these projects and express as a percentage. This indicated the percentage capital cost growth for the business. The source of the data was the project cost report produced by the project cost engineer. The initial target was to drive the capital growth

percentage to zero and thereafter continue towards negative percentages which indicate savings and increased profits or share of savings on GMP-type projects.

(4) *Quality.* At a given cut-off point every quarter, take the total number of non-compliances (NCs) for the previous 24 months and divide by the number of audits undertaken during the same period. Reduction in the number of NCs per audit can be a 'two-edged' sword as it may stifle innovation. Similarly, NCs should be used as a tool to recognise that something is not right and therefore highlights an opportunity to improve (the process or the person or both). A better PKPI for quality is often the amount of rework, both in engineering and in construction. Unfortunately, this information is not always readily available. With regard to engineering rework, a balance has to be achieved that does not result in individuals checking their own work to an inordinate degree as this will increase man-hour norms. Implementation also had to be handled with some sensitivity; however, the benefits of eliminating field rework were huge.

(5) *Tender costs.* The number of man-hours expended on all pre-qualifications and bids should be divided by the value of the contracts awarded and highlighted as the number of man-hours per £1000 of contracts awarded. The trend should be downwards; however, it has proved to be difficult to relate bid costs to contracts won because of the lag between bidding for and winning work.

(6) *Safety.* There are numerous indicators for measurement of safety performance; however, an indicator such as the accident frequency rate (AFR) which is derived from the number of lost time accidents (LTAs) per 100 000 hours worked on site on a yearly basis was a good indicator for our business. Agreed definitions were required for the LTA which tied in with the company statistics.

(7) *Resignations.* The intention was to retain and develop our core staff, therefore losing people was undesirable. This KPI was simply derived from the number of staff resignations in the quarter and expressed as a percentage of the total number of staff employed.

(8) *Qualifications.* Competency was and still is an ever-increasing issue with clients, and the higher the number of

'qualified' personnel (as opposed to only experienced personnel), the better the client's perception. This is not to detract from experienced personnel, but should be used to encourage personal development. This KPI was expressed as the number of staff who have achieved Engineering Council Registration divided by the total number of staff.

(9) *Financial.* This was initially derived from the profit variance at the end of the year compared to the budget figure. However, this ignored contribution to overhead, so a better KPI was the sum of the profit and overhead contribution at the end of the year divided by the sum of the budget figures at the beginning of the year.

(10) *Programme.* This KPI is a measure of how well we can meet the schedules agreed in the contract. This was determined by adding the sum of all the current projects' actual and forecast durations at the end of that quarter, and dividing by the contract project durations (in weeks) for the same period. Thus if the KPI is more than 100%, we are overrunning and missing milestones. Extensions of time need to be considered in the calculation.

Improvement phase

In hindsight the implementation of the KPIs was the easier phase of the work whereas implementing improvement initiatives required considerably more effort. Initially it was relatively easy to identify the areas that needed to be improved and the actions that needed to be taken. Many of these initiatives were originally instigated by the engineers and project managers; however after the 'baddies' had been addressed, we needed help, and best practice was sought out through a variety of sources:

- In-house throughout AMEC
- ACTIVE (achieving competitiveness through innovation and value enhancement)
- Construction Best Practice

These provided a useful source of material and ideas that have helped us improve significantly by a combination of improving our work processes and introducing new technology. As a consequence our project and business performance has been improved; however, there is still much to be done.

Lessons learned

It might appear from the previous sections that the improvements were brought about without any pain; however, this is far from the reality. Issues that had to be overcome were:

- Collection and manipulation of the data has proved very difficult even though the raw data were readily available
- Generating enthusiasm has been difficult with interest ranging from enthusiastic (the minority) through mildly interested to downright hostile (the majority)
- Persuading engineers to examine the way they did things was equivalent to challenging their manhood
- Senior management buy-in is vital to the success but hard to achieve
- Publication of financial data is extremely sensitive and strongly resisted
- Someone needed to be appointed to manage the monthly publication of the performance figures

Overall, the implementation has been a painful exercise but is now starting to show valuable returns. Some typical dummy trend charts are shown in the figures below.

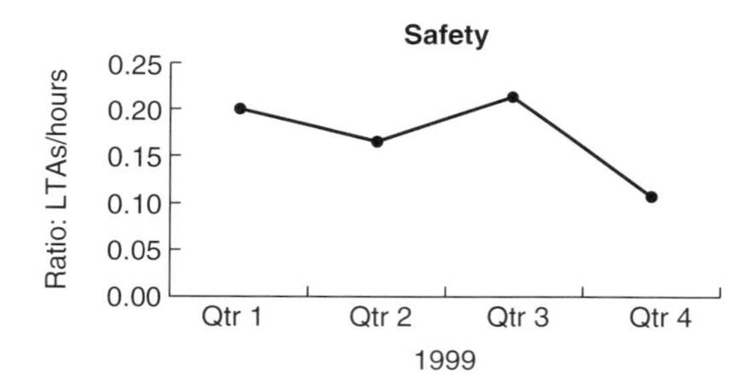

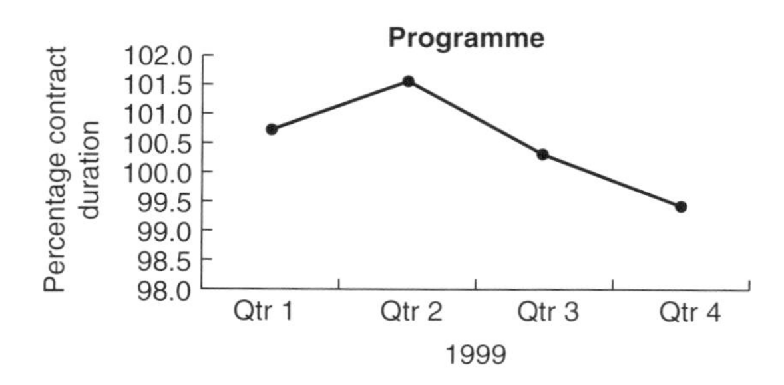

8.5.3 Case study three: Unlocking the secrets of successful organisations

Nicola Thompson, Business Improvement Manager, Miller Civil Engineering

Background

Over the last few years my experience in facilitating teams in benchmarking has led me to believe that there is a strong relationship between benchmarking and sustaining enthusiasm and support for change in the organisation. With the ever-increasing focus on both time and the bottom line, benchmarking has become a strategic necessity for teams and project champions to undertake.

Miller Civil Engineering (MCE) is a division of the Miller Group, a privately owned company that has an annual turnover in excess of £100 million. The division is a medium-sized civil engineering contractor based in Rugby with various offices scattered throughout the UK. It operates in all sectors of civil engineering including water, tunnelling, bridge and road building with work being primarily undertaken in the UK and has been registered with BSI as an ISO 9000 accredited company since 1992.

During the last seven years that I have spent within the civil engineering and construction sector, which has been spent in the most part within MCE, I have seen considerable change and less scepticism with regard to benchmarking. More importantly, I have observed the willingness of senior managers to learn from other industries as well as our competitors. This, they accept, is an essential part of their role in deciding future strategy.

My role as the business improvement manager within MCE has been to develop a strategy with other members of the management team, which would aid improvement, innovation and goal-setting that would ultimately affect the bottom line in a positive direction. My main responsibility is to ensure that my department meets its commitment to the objectives set out in the divisional and departmental strategy through KPIs designed to aid the monitoring of progress towards the company goals. Crucially, one of the most important roles I carry out is – in conjunction with the directors, operation and business managers – to set targets that will be used to monitor the effectiveness of initiatives used in various departments such as lean construction, just-in-time, focus groups, benchmarking, and the implementation of the Business Excellence Model. Additionally, I represent the interests of the company on forums such as the Major Contractors Group and the Best Practice Club.

Implementing continuous improvement in Miller Civil Engineering

Prior to 1995, MCE undertook very little work in the area of business improvement. Other than financial measures, no other metrics were in place. Certainly, the question of benchmarking between our competitors was raised but nothing happened: the sharing of our commercially sensitive material was often cited as the reason. Having spent some time in the manufacturing industry, this was totally alien to me. However, things changed and changed for the better. Continuous improvement teams were instigated, subject headings being based upon suggestions made by employees through the company suggestion scheme, audit results and internal meetings. The teams were cross-functional, led by a unit director and facilitated by myself. 'Continuous improvement' progress reports were completed which kept a tight rein on the team and set targets based upon the continuous improvement model (see figure below). However, although a good initial introduction into business improvement, they floundered. Primarily, the reasons for this were two fold: first, there was only one facilitator - myself - and too many continuous improvement teams for one person to facilitate; and second, project suggestions were deemed to be 'just updates to the company management system' and were not adding to the bottom line.

During this period, around 1997, the industry was, with the Latham (1994) and now Egan (Construction Industry Task Force, 1998) Reports, subject to much criticism and change. In addition to this, MCE clients were also shifting their attention from employing

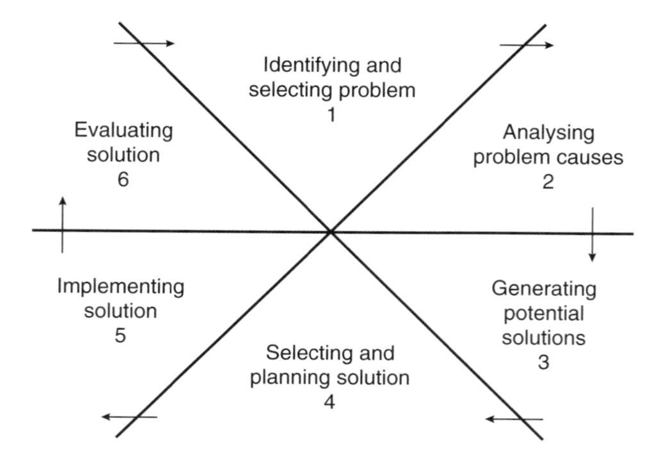

The continuous improvement model

a contractor who had ISO 9000 or equivalent to one that could demonstrate and implement change within their organisation. Coupled with this, I was also starting to build up a number of contacts within our own industry and outside, that were willing to share information. All rather fortuitous as we were now experiencing problems with some of our processes that were clearly affecting our bottom line and we could visibly benefit from learning from others – the essence of benchmarking.

As a result, MCE joined a number of forums including the Midland Construction Forum in 1994, the Best Practice Club in 1997 and the Major Contractors Group in 1998. The membership of these forums varied. The Midland Construction Forum generally had a membership audience from our own industry sector with support from academia; the Best Practice Club a network of members from all parts of industry that were, and still are, administered by an external body based in Bedfordshire. The advantage of the latter was that its membership list included a number of industries of which best practice was recognised and could be sought, whilst at the same time observing confidentiality with the 'administrators' undertaking the introductions, workshops and case study days. The final forum that MCE joined on behalf of the Miller Group was the Major Contractors Group (MCG). This was a specific benchmarking club set up as a clear demonstration of MCG's positive response to the Egan recommendations, with participants from all the major contractors within the UK.

The benefits that MCE perceived could be gained from joining and participating in such forums were:

(1) To allow MCE to compare its performance in key areas
(2) To demonstrate to its clients and Government forces that it was responding positively to changes within the industry
(3) To provide statistical information on how MCE was performing compared to other sectors of the construction industry with a view to identifying best practice

Coupled with this, by joining such forums it paved the way for further avenues and doors for benchmarking initiatives. By joining the Midlands Construction Forum and inviting academia, it enabled the forum to apply for grants, for example from the Engineering and Physical Sciences Research Council (ESPRC), to help facilitate and drive improvement and identify best practice in specific areas. As MCE is a member of this forum, they could reap the benefits of this if they so wished. An example of such a project is the Culture of Quality which is being undertaken with the aid of an ESPRC grant with four participants from the Midlands Construction Forum taking part, one of which is MCE with Birmingham University acting as the academic partners. The aim of this project was and still is, and the benefits that MCE could perceive from their involvement were:

- The identification of impediments to the development of a culture of quality in the organisation
- The development and application of methods (tools) to promote a culture of quality in the organisation
- The raising of awareness of quality issues among the company's staff
- The facility for key employees of participating companies to take part in discussion groups, to raise awareness of the cultural dimensions of quality management
- Development of appropriate benchmarks for specific processes, allowing the company to compare their performance with that of others in the study
- The application of tried and tested techniques to profile the culture of the organisation
- The opportunity to participate in the testing of new techniques for improvement to be developed during the project

Whilst MCE were actively immersing themselves in these forums, there were also moves afoot in other areas. Having learnt from their previous experience with continuous improvement teams, a Business Improvement Forum was instigated chaired by a director facilitated by myself with four other members of at least middle management seniority. However, this time, to ensure that only value-adding ideas were taken forward, that projects were managed in a controlled environment, and to encourage top-down, bottom-up ideas, all the output from the attendance at forums, the company suggestion scheme, etc. would be fed into the forum. From here business improvement teams would be instigated – cross-functional, cross-unit and led by a 'champion' whose responsibility it would be to feed back progress with the initiative to the Forum. The benefits from this, MCE saw as twofold:

(1) It would act as a central focal point which would encourage all levels to get involved but would have a senior person as chair who could make any necessary decisions and provide a feedback loop to the MCE Board

(2) It would ensure that MCE was responding to the best practice established from membership of the forums

Using KPIs in Miller Civil Engineering

In addition to the foregoing, MCE instigated a number of internal KPIs (see table below) which again would encourage 'best practice' by internally benchmarking between divisions and units. The KPIs, which might have been considered too simplistic by more advanced companies, were for MCE, a starting point whereby 'quick wins' could be obtained. This was considered to be essential if we were to keep the momentum going within the organisation. We had to remember that this was a complete shift from the previous culture and we recognised that the barriers that exist with external benchmarking clubs, i.e. the sharing of information, would also be present within our own units within the division. The MCE approach was in line with the steering group involved in the MCG, and the implementation of the Egan KPIs is shown in the table below. Instead of trying to implement all ten of the Egan KIPs, the steering group have taken on board just five, for the same reasons already cited. In order to test the methodology that they have chosen, four of these five were implemented (see table below).

Egan KIPs	**Internal KPIs**
(1) Cost predictability	(1) Number of employee suggestions implemented
(2) Time predictability	(2) Number of continuous improvement projects completed
(3) Defects 1 and 2	(3) Number of ISO 9001 non-conformities
(4) Accidents	(4) Plant breakdown
	(5) Number of customer complaints
	(6) Energy consumption rate
	(7) Number of environmental reportable incidents

MCE have also decided to apply the same KPIs that they instigate and the rigours of the Business Excellence Model further on down the supply chain. This, MCE have envisaged, will enable them to measure the performance of their vendors by both a self-assessment methodology and external validation on their vendors.

In order for MCE to participate in the external forums, one person has chiefly been involved, namely myself. Although this would be considered by some as 'putting all your eggs into one basket', MCE have found it beneficial as it provides a focal point and a person whose responsibility it is to again give feedback to the Business Improvement Forum. The resources that are being used for the focus groups with, for example, the Culture of Quality project with the Midlands Construction Forum and the University of Birmingham and the internal Business Improvement Forum/ Teams, are considered to be cost-effective as the following benefits ensue:

- Encourages a good cross-section of personnel to be actively involved in the way that MCE may run their business in the future
- Encourages sustained growth rather than one-off 'quick fixes'

One important point to note is that the majority of personnel, approximately 75%, and certainly all the personnel involved in the projects, are trained in continuous improvement tools and techniques. The ethos for improving and the benefits that benchmarking can bring are now set in stone and are driven from the top down. The facilitator is experienced and trained in this area.

What benefits have Miller Civil Engineering gained from benchmarking?

Since participating in various benchmarking clubs and forums, MCE could cite a number of examples whereby such involvement has either reiterated areas for improvement that MCE thought was already a weak area, or given the organisation an extra push and expertise to instigate new tools and techniques for improvement. It has also given us the confidence and trust to share information, knowing that it will not endanger commercial confidentiality. The table shown below highlights some of the benchmarking projects MCE have been involved in, the forum and the benefits they feel they have gained.

Project	Medium	Benefits
Egan KPIs	MCG	• Use the same KPIs to benchmark our vendors' performance • Use the data that have been submitted to internally benchmark within the individual units in the division • Benchmark against the industry best practice
The procurement process	Midlands Construction Forum	• Highlighted deficiences within MCE procurement process which has resulted in positive cost savings of 70%
The Business Excellence Model		• The use of a self-assessment tool which will be used to good effect further down the supply chain
Culture of Quality		• Increase in quality awareness and the varying cultural dimensions within different functions and levels within MCE
KPIs	Instigated by MCE	• Reduction in audit findings both internally and externally over the last three years • Fewer customer complaints • Increase in productivity in key areas, for example plant • Decrease in environmental incidents

Critical success factors for benchmarking and establishing best practice

These can be summarised as follows:

(1) The longer it takes for a team to complete a benchmarking project, the more difficult it becomes to sustain enthusiasm and support for change within the organisation
(2) Primary best-in-class company research and establishing industry information sharing relationships can often consume over 50% of a typical benchmarking project timeline
(3) Become members of specific clubs and forums that promote

best practice and best-in-class as it enables better alignment and focus of members and the benchmarking team

(4) Ensure that the club is right for you and the key areas that are to be benchmarked are integral to your business and will have some effect on the bottom line

(5) Remember it takes time to build up the rapport and more importantly the trust between members. Use standard Codes of Conduct such as the European Code of Conduct

All of the above ensure that:

- there is commitment from the top
- there is total belief in the benefits
- the personnel are skilled and trained and have the enthusiasm and authority to submit and share data and put actions in place to help make your company – *Best-in-Class*

8.5.4 Case study four: Benchmarking for best practice in the construction of Hilton International

Michael Collini, Project Manager (Technical Services – Europe, Middle East and Africa), Hilton International

Introduction

Hilton International is the main subsidiary of Hilton Group plc (formerly Ladbroke plc) a Footsie 100 company that also boasts the leading betting and gaming brand name of Ladbrokes and the LivingWell Health and Fitness Clubs. Hilton Group produced a turnover in 1999 of some £4.3 billion with profit before interest and taxation of £305 million. Hilton International is present with 217 hotels in some 55 countries and their hotels are a mixture of owned, leased and managed properties, a common feature of the fragmented nature of a hotel's property portfolio. In 1999, Hilton and its hotel owners invested £350 million in capital for hotel construction, renovation and extension works. The planned expenditure for the year 2000 is £338 million.

My role as a project manager deals primarily with the overall responsibility for the management of complex refurbishment projects. I represent Hilton, a corporate organisation in which property and construction is not a core activity, in the refurbishment process. Internally, this involves the co-ordination of Hilton personnel on our operational side, an important liaison as we rarely close our hotel operation during refurbishment works – an added challenge. Externally, I manage the project team and oversee the contractor. I am a chartered quantity surveyor by profession and, before my current role, I worked in Milan, Italy, for an international construction consultancy. This provided my first exposure to a foreign market and the basis for my current role with Hilton.

Our technical services department, now decentralised, has a small core of construction professionals – architects, services engineers, cost consultants, project managers and interior designers. We oversee all new hotel development work and refurbishment projects over £500 000 in value. Due to the extent of the geographical area we cover much of the work is outsourced to construction consultants. What we are constantly searching for is to discover best practice that will allow the construction of Hilton Hotels to proceed more speedily and efficiently than has been the case in the past. As such, at Hilton Hotels we are entirely supportive of the aims that Sir John Egan proposes in *Rethinking Construction* (Construction Industry Task Force, 1998).

The background prior to improvement

The company's hotel refurbishment projects were notorious for exceeding both time and budget. Capital expenditure had been badly managed. Priorities were given to the refurbishment of public areas and guest rooms as these supported the necessary return on investment criteria. The back-of-house areas and the services installations only featured where mandatory requirements dictated. Reactive maintenance was the norm. All this is perhaps an indication of an organisation where property is a non-core function. The need for a professional approach to project management had arrived.

The approach adopted

The approach we take to facilitate the way we operate over three continents is one of simplicity and flexibility. In the UK we take for granted that monthly project meetings will be held, that agendas will be prepared beforehand, that minutes will be issued thereafter and so forth. These basic procedures are somewhat alien to many countries. However, a formal regimental approach is not the answer. We believe that basic concepts of standard practice can be introduced simply and informally. It is important to understand the different cultures that exist in the areas we operate in and to respect these differences. A basic grasp of the local language is useful; even simple greetings assist in helping to break down initial barriers. Our objective is to establish consistencies in practice over a multi-cultural perspective.

What has been achieved – some examples of best practice procedures

Our involvement in a project usually commences with the preparation of an investment plan for board approval. This involves developing a long-term investment strategy for the hotel in conjunction with our operational specialists. The opportunity to reposition a hotel in its current market will be studied. The driver will be the profit-generating elements of a project that justify the return on investment necessary. However, a co-ordinated approach is essential so that the mandatory requirements (upgrade to satisfy current legislation) and the maintenance aspects of a project are properly considered.

An interior designer will often be appointed to develop some initial design proposals and an independent cost consultant will always be appointed to establish the project budget. The cost consultant must be independent from the design team in order to provide impartial advice. This will then develop to the necessary cost control during the design process. The role of an independent cost consultant is not one that is common to many countries and is therefore often provided by an 'off-shore' consultant. In respect of project costs, we have also developed and are continually updating our hotel cost database. Total project costs are analysed using simple cost parameters such as cost per room and cost per square metre. Partial project costs such as cost per cover for restaurant works or cost per person for meeting room or ballroom refurbishment have also been prepared. A detailed cost database for guest room refurbishment has also been developed. Whilst the spread of costs differs enormously due to the varying scope, the database provides useful comparisons at individual unit rate level.

A project budget form (PBF) will be prepared that forms the basis of the request for capital. The total project budget is to include the cost of land, construction, FF&E (fittings, furnishings and equipment), professional fees, statutory charges, contingencies and head office project-related expenses. The project budget form is then incorporated into an authority for expenditure (AFE) form. Approval is required from the main plc board for major project expenditure. This initiates the design and tender phase of the project. A final administrative procedure, referred to as the authority to proceed (ATP) form, is necessary before signing the main construction contracts. This is essentially a simple check-list to ensure that simple but fundamental issues are in place, e.g. has planning permission been obtained?, is the design complete?, etc.

We have also developed a Project Procedure Manual. This acts as a handbook of step-by-step procedures that takes one through a project cycle. Whilst it does not provide a textbook solution to each individual project problem, it sets out a process that installs good management and discipline from the outset.

In respect of good practice from a technical point of view, we have over the years developed our 'Hilton Technical Guidelines'. These contain detailed specifications of our project requirements and technical standards that form the basis of the brief to the design team.

The importance of adequate preparation for a major project is one we recognise. Planning approval procedures vary considerably from country to country. Although the process is often the same, the time-scales differ significantly. We rely heavily on the local expertise of our appointed architect to guide us through this process. This leads me to the appointment of the Project Team. We seek to have a balance of local and international skills. As I mentioned, a local architect or planning consultant is essential not only to deal with the local authorities regarding the approval process but to advise on local statutory requirements which again differ considerably across the areas in which we operate. The 'off-shore' or international project team roles relate to those that are somewhat specialist in nature and not often sourced locally. Such examples include the interior designer, the lighting consultant, the acoustic consultant and the cost consultant. We have a core network of approved specialists in these fields with whom we have worked for some years, which also brings the added advantage of mutual understanding. In appointing any member of the project team, we look for the right attitude and enthusiasm which is backed up by solid hotel refurbishment experience.

The importance of adequate preparation is also reflected in the design process. Our objective is to develop as complete a design as possible before the issue of tender documentation. During the design phase of a project, we will study realistic construction time-scales based on the actual scope of work. In the past, construction programmes have often been established based on the hotel's operational requirements, i.e. sell-out dates due to major event, and the programme has often been unrealistically reduced to meet such requirements.

In procuring construction contracts, we value the importance of a fixed lump sum price despite the inherent nature of refurbishment projects. The most common procurement method we adopt is a traditional one using a main contractor but we also consider employing trades on a separate basis where this is favoured locally. We recognise the different legal frameworks that exist in the countries in which we operate. As a result, we have established common contractual principles that are then incorporated into a local form of contract rather than imposing a standard form of contract.

As part of an ongoing learning process, we aim to educate and train our operational management in the project process. Hotel general managers and chief engineers do not come from a construction background, yet they play an important daily role. The interface between site and hotel is critical to ensure the correct balance between adequate site progress and the continuance of hotel trading. The safety and security of our hotel guests is, however, of prime importance and not to be compromised.

During the construction phase, we normally arrange formal monthly project meetings between the hotel, project team and contractor to review strategic project issues. We have developed standard monthly procedures for internal reporting that are simple and user-friendly.

Upon completion of a refurbishment, we have established a post-project review. This is held with the project team and contractor to encourage constructive criticism and comments that hopefully benefit and improve us as a client organisation in the project process.

Lessons learned

Notwithstanding all the good will and intention, there have been problems and some painful lessons have been learnt. For example, following a significant cost overrun on a major refurbishment, it is now our policy to carry out extensive intrusive condition surveys for any major refurbishment project. Similarly, it is no longer acceptable to rely on as-built drawings of properties that are over 30 years old. We have also decided to recruit a full-time on-site Hilton project representative. Our hotel managers are not construction people. Their task is to manage the operation of a hotel. We require an extra independent 'watchdog' to supervise the project team and the contractor in order to 'flag-up' early warnings of potential problems.

What we are seeking to do as continual improvement

From a technical point of view, we are preparing a mock-up bathroom based on pre-fabricated bathroom modules. These will be delivered in pre-finished elements (two floor and ceiling pieces, eight wall elements) and installed in two days. Whilst the capital cost is very comparable with a traditional installation, the anticipated saving in time will have a major impact on our rooms revenue. The quality of the finished product will also be much higher as a result of the factory conditions in which it is produced.

We are also currently constructing four prototype 'rooms of the future'. These have been developed following extensive marketing research of guest requirements and incorporate many futuristic features such as a fingerprint room access system and voice-activated controls to baths and basins

In order to improve our organisation in the project process, we are preparing a handbook of standard definitions and common terminology of project-related issues. This is essential in a global organisation where construction is not a core business and where languages, cultures, legal systems, etc. differ from country to country. This handbook will define, for example, what is included in the project costs; what items are covered by the contingency sum; what we mean by liquidated damages. We welcome the recommendations of the Egan Report and are seeking to introduce a partnering style arrangement with our consultants and main suppliers.

8.5.5 Case study five: Using the Barhale 2001 Model to create continuous improvement

Mark Evans, Business Excellence Manager, Barhale Construction plc

Background

Barhale Construction plc is a modern provider of construction services to utility and public sector clients. Having a natural work philosophy aimed at providing a client-focused service, the company has thrived in the post Latham/Egan era. It now derives much of its turnover from long-term programmes of work within alliance with blue-chip utility companies. Barhale's range of services includes:

- Water and wastewater network services
- Telecommunication networks
- Specialist mechanised tunnelling and shaft-sinking
- River and canal engineering
- General civil engineering that includes the building and repair of roads and bridges

As will be described below, the development of a wide portfolio of work has been achieved by the commitment of the company to continuous investment in plant and, importantly, people. Crucially, an overriding concern at Barhale has been to ensure that, even though the range and value of work carried out has steadily increased, the company's reputation for quality, reliability, safety and integrity remains intact. For this reason, this case study will explain the importance of looking to our people to provide the expertise and dedication which, we believe, has made it possible to create the opportunities for building on our existing client network and seeking new contracts. As will be explained, the Barhale 2001 Model has been a vital part of the improvement activities being carried out in this organisation.

The influence of customers

In modern business, it is accepted that customer satisfaction is an essential part of ensuring continued success. At Barhale Construction we stress that this is what we must continuously strive to achieve. The fact that we have been dedicated to use of performance indicators to measure our ability to achieve targets is a result of the fact that most of our work comes from clients who have been at the forefront of using such methods Any organisation seeking to procure work from such clients will quickly discover that this approach is pretty much a prerequisite. Our commitment to ensure that we deliver continuous improvement to achieve customer satisfaction was embodied in our challenging improvement programme 'Barhale 2001'.

The development of the Barhale approach to excellence

Our experience of working with customers in recent years has been that they expect providers of products or services to be proactive in showing how they will be able to give 'more for less'. The Egan Report argued forcibly that the construction industry must do exactly this if it is to improve. As many of the customers we work for are involved in the provision of utility services, they are in the position of having to show increasing returns to shareholders and at the same time improve the quality of service to customers. As a good many of these used to be in the public sector, there has been some very rapid learning of how to compete effectively in the open market. Additionally, there is the added dimension of the need to satisfy regulators, such as OFWAT in the case of regional water authorities. In order to explain the development of performance indicators at Barhale, the experience of working with British Telecom (BT) will be used. As we have found, the need to provide definitive measures of performance for this customer has proved to be invaluable in benchmarking other areas of the business not connected to ductwork and cable laying.

The telecommunications market is extremely competitive with a considerable number of competitors vying for business in what was, until relatively recently, a monopoly. BT has risen to these challenges and its overall performance today is not just better than previously, but it is considerably more visible. A cost involved in the provision of telephone services by BT is for the laying of ducts and cables. Those tendering for these work programmes were expected to show how they could achieve targets such as completion to time and cost, but in addition to demonstrate a commitment to 'adding value' to all stages of the process.

Like many other organisations in recent years, Barhale implemented a quality assurance system. Whilst the use of QA is beneficial in demonstrating conformance to stated requirements, we recognised the need to improve all aspects of our business – including, in particular, areas not covered by traditional QA (e.g. leadership and people issues and a culture of continuous improvement). Our overall objective became the development of mechanisms to understand, define and measure our performance and to use these 'smart' data to drive improvements in the business. We strongly feel that these approaches should be established as part of the way we run our business, rather than just a 'bolt-on' for key clients. Our framework for managing our total approach is the EFQM Excellence Model. This establishes the key principles of Enablers causing Results, and Results driving Enabler improvements. The opportunities to benchmark both Enablers and Results are exciting and are facilitated by the 'language' of the model.

It was decided by the board that steering groups should be set up in each of the Enabler criteria of the EFQM Excellence Model:

- Leadership
- Policy and strategy
- People
- Partnerships and resources
- Processes

The result of this decision was the Barhale 2001 Model. As the diagram below shows, a number of teams were established to consider improvements that most directly impact on our ability to achieve success in the Results criteria (especially in customer satisfaction). Whilst the work of all these teams is crucial, the progress of those working on process improvements is of particular importance as the improvements generated are often 'quick fixes' that can give a real sense of achievement. Overall, there are seven improvement groups; four of these focus on the issue of people and two on processes. We have learned that a fundamental part of introducing improvement is the necessity to involve people and from an early stage. The need to treat people as the most valuable resource has become something of a cliché in industry and we recognise the need to establish some enabling activities to ensure that we get the most from our people. In support of this, there needs to be a visible commitment by the organisation to ensuring

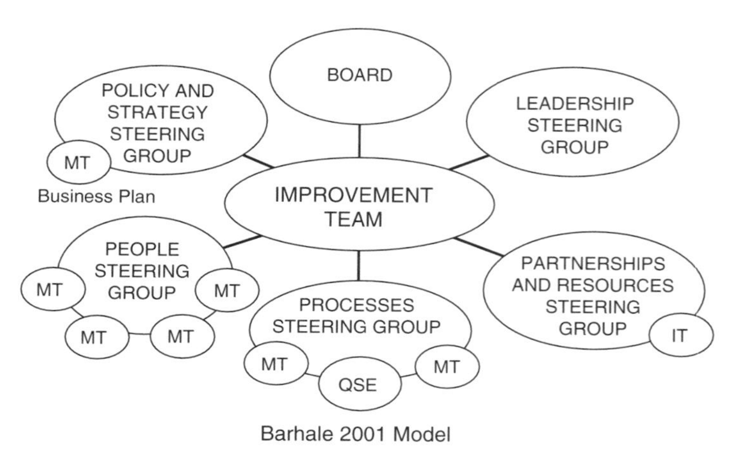

Barhale 2001 Model

MT = Management Team
QSE = Quality Steering Executive

that people are treated in a way that guarantees their efforts are recognised and rewarded. And just as importantly, we realise that people at all levels need to be held accountable for the achievement of targets. Our goal, therefore, is to provide our people with every opportunity to contribute and to develop their work and themselves, whilst at the same time keeping a strong focus on what is important to the business and what is actually being delivered.

From the results of an early self-assessment against the EFQM Excellence Model, the company introduced the Investors in People Standard to guide our human resources strategy. This has subsequently yielded great improvements to the development and satisfaction of our people.

Whilst the dedication to measurement is clearly a good thing, it can lead to an overabundance of data which may be confusing and overwhelming for those trying to take decisions. An important principle embodied in the Barhale 2001 Model is that those who are involved in day-to-day operations are those who are best placed to measure and improve the way that they are carried out. Unless the measurements carried out are honest, it will be difficult for effective solutions to be established.

Ownership is, we would argue, the most important aspect of improvement efforts. We want our people to be genuinely committed to the concept of improvement and to take ownership of the solutions they generate. This requires a high degree of empowerment and trust from the senior management. Through this philosophy, we believe that the chances of success are higher than if they had been 'imposed'.

We wanted to deploy Barhale 2001 within existing contracts, including that with BT. One of our key drivers was a better relationship with the customer, based on an improved mutual understanding of each other's needs. For instance, at the commencement of the contract - some four years ago - a primary means of communication was by corrective action requests. All parties felt that there was an opportunity to take a more proactive approach to working together (and ultimately, of course, to remove corrective actions altogether). This objective required BT and Barhale to actively engage in ongoing dialogue. One benefit of this approach is that, by better anticipating each other's work requirements, it is possible to have the right resources in the right place. Sometimes this requires compromise by each side but this is seen as a healthy demonstration of the commitment to work cooperatively.

Using performance indicators to drive improvement

At Barhale we concentrated on developing effective performance measures to show the impact of improvement in key areas on each project. To meet our objectives, we will need to drive all the things that will deliver success by regularly setting targets and measuring performance against them. Using key performance indicators (KPIs) we measure in the three areas that combine to form a full picture of our business:

(1) The EFQM Excellence Model
(2) Headline indicators
(3) Operational indicators

These three areas are far from disparate, in fact they all fit together. The Operational KPIs feed upwards into the Headline KPIs and all of these can be mapped against the EFQM Excellence Model which, as a holistic framework, covers all aspects of the business. This in turn will identify the gaps that are critical if we are to become truly excellent.

At the top of the KPI pyramid we use the Headline KPIs promoted by the Construction Best Practice Programme (CBPP). These are widely adopted and will allow us to look outside our businesses and compare within the industry.

Feeding up to these are Operational KPIs from both the CBPP and those designed by us with our client to match their specific programmes. These allow managers to focus on specific areas for improvement. The following list shows some of the operational areas that we measured, every three months, on a large tunnelled water supply project undertaken with Thames Water:

(1) Safety record
(2) Programme control
(3) Financial control
(4) Public relations
(5) Changes
(6) Disputes
(7) Correspondence
(8) Risk management
(9) Value engineering
(10) Managing change
(11) Damage and loss incidents
(12) Staff ability
(13) Workforce ability

(14) Personnel turnover
(15) Absenteeism
(16) Project team's values
(17) Client - contractor relationship
(18) Significant events

With the exception of staff ability, workforce ability and significant events, each of these is given a score out of five (5 = excellent; 4 = good; 3 = Average; 2 = below average; 1 = poor). As a consequence, the project could score a maximum of 85. The result of using this scoring system has been that we are more active in ensuring we can identify areas where improvement is urgently required. More importantly, we are able to use the method to encourage employees involved in each project, using them to constantly search for ways to improve the scores.

One of the most useful developments of using this scoring system is that we are able to use the data collected to compare ourselves internally and externally. Managers are keen to learn the way in which projects which score highly have been able to do so. As we would argue, those primarily involved are using the method to assist themselves in their own activities, not because it has been imposed from head office.

From our experience there are some key practical points to be aware of in the implementation of KPIs:

(1) High-level commitment is essential
(2) Avoid volume overload - keep it simple and wherever possible use existing data
(3) Indicators will continue to develop as expertise in using them increases
(4) It is important to dispel concerns people may have about them being just a ranking tool: it is improvement that counts and we must encourage a high level of honest reporting
(5) Know your audience. For example, a managing director will want a certain type of information that will allow them to market the organisation. However, a site manager will require data that are current, and therefore, will enable them to understand what has occurred today and consider how to create opportunities for improvement tomorrow.

Conclusion

There is no complacency at Barhale Construction. We have made improvements at the same time as generating more business, but we acknowledge that markets will continue to be influenced by clients who are increasingly more demanding in what they expect from their suppliers. We realise that our clients are themselves being driven more and more by their own customers and stakeholders and that the best approach must be to work together towards common business objectives. This requires trust and openness. By using performance indicators, we are confident that our customers and partners will better understand our commitment to improvement. After all, you can't argue with transparent performance data – especially when the performance is improving all the time. We would recommend that other organisations take on board the principles of business excellence and start managing by fact.

8.5.6 Case study six: The road to world class – learning lessons from the 'best'

Hamish Robertson, Quality College Manager, Morrison plc

Utilising people – the cornerstone to our approach

Morrison plc is an Edinburgh-based company which has, we believe, a long tradition of being dedicated to improving the quality of its product and services provided to clients. In the late 1980s we installed quality assurance with a view to increasing consistency. However, whilst BS 5750/ISO 9000 (BSI, 1994) was, and is, effective in ensuring that quality was achieved by the use of documented procedures, we needed an alternative approach that captured the knowledge and expertise of our employees. What we wanted to do was to create a culture in Morrison that made dedication to excellence something that everyone in the company truly believed in and was involved with. In essence, people became the focus of improvement activities in Morrison. In order to assist employees to be able to more actively contribute to improvement activities, senior management was given two key responsibilities:

(1) To ensure that employees were trained and developed towards providing them with, as a minimum, sufficient skills and confidence to do their best
(2) To create a culture that would support everyone's efforts in providing customer delight

The former was supported by the achievement of the standard required for recognition as an Investor in People. The latter required a constant search for improvement in existing ways of working. Of even greater importance was a dedication to providing training to our employees that would allow them to fulfil their potential. As we stressed to all of our employees, primary responsibility for training rests with them. We did not want to force people to attend training programmes; to do this, we believed, would simply be a waste of their time and ours; worse still, it may cause resentment. However, this approach does not remove managers from their role as facilitators of improvement. We encourage them to be constantly aware of suitable or appropriate training packages or programmes that will allow our employees to improve their skills in crucial areas. The following diagram shows what we call the Morrison Training Cycle and summarises the main elements of what we are trying to achieve.

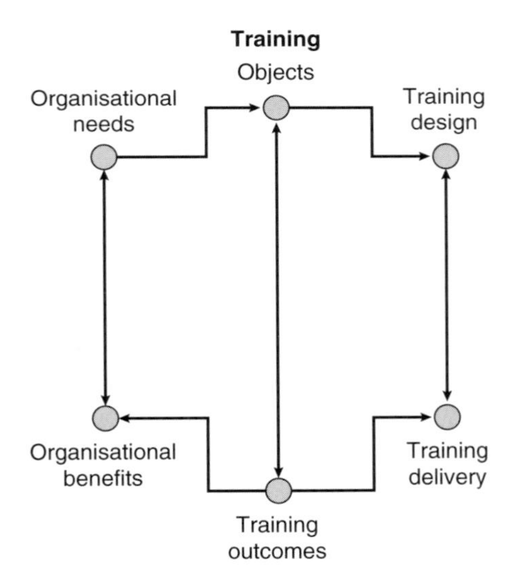

As a consequence of using this cycle, we have adopted the following slogan:

> Working – and training – together in partnership, we will achieve our goal to be world class

As the next section describes, in order to become world class, we have had to be prepared to benchmark ourselves against organisations that are recognised as among the best.

Benchmarking as the basis for becoming a world-class organisation

One of the things that is important in the quest for excellence is the ability to measure the effects of implementing improvement initiatives. As senior managers at Morrison accepted, the only way to achieve this was by comparing our performance relative to others considered to be 'better' than us. In the first instance, this involved comparison with other contractors, a method known as competitive benchmarking. However, as will frequently be discovered when this type of benchmarking is used, there are two distinct problems:

(1) The fact that the organisations against which comparison is being made are in the same market tends to mean that they will not divulge to direct competitors how they have created opportunities for improvement
(2) Whilst competitive organisations may be regarded as being very good within their own sector, in comparison to accepted world class, they may not be acknowledged as such

Therefore, senior management decided that on the basis of the group's mission statement – 'We aim to deliver world-class customer service and value through our innovative, quality-driven approach' – we needed to compare ourselves against the best in the world. However, in order to do this, a methodology was needed that would allow us to provide what could form the basis of objective measurement. As we became aware, various types of methods existed to achieve this. Whilst these provided inspiration, we wanted to develop a form of measurement that reflected an approach that employees at Morrison would recognise as being 'home-grown', and therefore, feel they owned. The culmination of this process was the 'Balanced Business Scorecard', the means by which Morrison can ensure that it is making progress towards becoming world class. The fact that 18 key performance measures were incorporated was a useful device which allowed us to draw an analogy with a scorecard used in golf; the scoring used reflected the desire to attempt to achieve par in each of the following areas:

(1) *Safety* – to engender in Morrison a world-class culture and performance. In order to do this we measure lost hours, causes of accidents, the frequency of certain accidents and the amount of health and safety training that is provided to our employees.
(2) *Strategy and policy* – to provide demonstrable focus on

becoming world class. This, we accept, is an essential area that we must constantly re-evaluate in order to provide the focus that will allow us to attain world-class status.

(3) *Efficiency* – the aim here is to both increase efficiency and reduce waste. By doing this, we will be able to provide more for the same amount of money; something that will enable us to improve cost savings to us and the client.

(4) *Supply chain management* – we need to be able to more accurately measure and improve the value of the service and products that are provided to us by suppliers and sub-contractors.

(5) *IT capability* – our employees need to be provided with the best systems and training that, we believe, will allow them to provide our customers with a level of service that is perceived to be world class.

(6) *Five-star sites* – the quality of what we do on our sites is, we stress, critical to gaining competitive advantage.

(7) *Customer satisfaction* – this, of course, is the aim of any organisation; to ensure that what the customers receive matches their expectations, and better still, surpasses them. This measure is based upon feedback from interviews and questionnaires and allows us to identify strengths and weaknesses so that we can focus our attentions appropriately.

(8) *Customer value* – this is a measure that companies such as Pepsi and McDonalds use to assess their ability to deliver high quality at a cost which is felt by customers to be reasonable. We want our customers to champion us by repeat business and recommendation as providing high value at acceptable cost.

(9) *Impact on society and the environment* – this is something that all companies wish to demonstrate they are doing; providing their services or products in a way that contributes to society and the environment in a way that is non-harmful.

(10) *Partnering* – as most organisations have discovered in business relationships, working together towards mutually identified goals will always be more effective than trying to achieve objectives where conflict ensues. As a consequence, we now measure our ability to carry out business in ways that allow us to co-operate and form alliances.

(11) *Employee satisfaction* – this is done by allowing every employee to provide us with measures of their morale, achievement, contribution and suggestions for personal and professional improvement.

(12) *Training and development* – here the aim is to measure our ability to provide skills that meet business and personal goals. Like the previous measure, we use methods of assessment that enable us to monitor training (see below for an example of this).
(13) *Teamwork and leadership* – we believe that in order to become world class it is essential that we are able to foster a climate in which teams are led by those who have the ability to inspire. This measure monitors the effectiveness of those who have been given responsibility to lead teams.
(14) *Innovation* – the aim of this measure is to encourage creativity. Thus, we measure how effective we have been in creating a culture that allows people to propose and implement new ideas.
(15) *Earnings per share growth* – as with any public-owned company, we have a duty to our shareholders to provide a healthy dividend. Even though this may seem like a fundamental measure used in accountancy, it provides a gauge of how, by creating opportunities for improvement, we are able to meet the expectations of investors.
(16) *Profitability* – like the previous measure, we want to show to potential investors that as well as giving our customers a world-class service and providing valuable opportunities to our employees, we can return consistently good profits. This measure allows us to show that the quest to become world class benefits all concerned.
(17) *Risk management* – this measure gives us an assessment of the potential risk to our operations of particular decisions. As we accept, whilst the intention is to allow radical solutions to traditional problems, we should not put the business at undue risk.
(18) *Shareholder added value* – like the other financial measures above, we want to show how effective we have been in creating additional return to those who invest in Morrison.

These 18 measures provide the means by which to benchmark our achievement of progress in becoming world class. In effect, they give us an accurate assessment of the way in which continuous improvement is being effectively implemented. Moreover, as we have become increasingly aware, clients are much more demanding in their expectations of the ability of supply organisations to demonstrate how they use key performance measures. Using the balanced scorecard allows us to do exactly that.

Measuring up against the best – putting the Morrison approach into action

As all employees are informed, the objective of becoming a world-class company is to enable us to compete more effectively, and as a result, we stress, to make everyone's working lives more fulfilling (and, of course, secure). In order to emphasise this message, literature is supplied which provides examples of how world-class companies have achieved radical improvement in certain areas of their business. The intention is, we explain, to inspire our employees to think about what they could do which might enable us to implement initiatives which are similarly spectacular. Some of the examples that are typically used are:

- *Efficiency* – the example of Porsche being able to reduce the time taken to produce a high-performance car from six weeks in 1991 to three days in 1998. In addition, it is explained, errors have been reduced by 75%. Even more significantly, Toyota housebuilders are cited as being able to carry out 80% of construction in a factory environment in which standards of quality are far higher
- *Partnering* – the example of Motorola which has cut production costs by up to 35%, reduced cycle time by 99% and, significantly, has been able to cut their order lead time from 8.75 days to 15 minutes whilst at the same time increasing market share by 15%.
- *Supply chain management* – the fact that by being able to deal more effectively with suppliers, it is easier to make improvements. For instance, whereas the average car manufacturer in Europe deals with 4700 suppliers, in Toyota this figure is only 400.
- *Customer satisfaction and development* – a number of organisations (e.g. Daimler Benz), are used to show the way in which customers become the 'touchstone' of success in becoming excellent.
- *Training and development* – as Morrison states with respect to this area: '... many Japanese companies invest between 11 and 15 days per employee in training ... our aim is to be up with the best in the world'. However, a qualitative measure is also essential, the benefits of which are captured through quarterly employee surveys.

As quickly became obvious in making progress towards the goal of becoming world class, the most essential element in attaining success is what EFQM call the 'key resource' – people. However, whilst the theory of ensuring people are actively involved in improvement efforts is all well and good, actual practice can prove to be somewhat more difficult. Time and effort can be considerable, and there has been the additional problem of convincing people that there is an alternative to the traditional short-term and contractually orientated way of doing business. In effect, the quest towards becoming world class became one of winning the hearts and minds of those who would be part of the 'cultural revolution'.

In order to do this, a number of people within Morrison became facilitators whose task is to assist others to more effectively carry out, among other things, the following:

- Measure the effects of improvement efforts on day-to-day activities
- Use problem-solving techniques
- Communicate to each other in such a way as to exchange vital information
- Make presentations

As is continually stressed, whereas in the past these activities might have been seen as being superfluous to working in a contracting organisation, now they are regarded as being key business skills. The objective is, therefore, that the desire to engage in improvement becomes 'second nature'. As has been described previously, one of the benchmarks of an excellent organisation is the amount of time dedicated to training. The average benchmark set by Japanese organisations is 15 days. Whilst Morrison has not beaten this – currently dedicating 9.5 days – it is significantly more than the four days reported by DTI as being the average for the UK.

A vital part of the effort to ensure that people are willing to be involved in dedicating time and effort to the quest to become world class is that they are recognised. This does not just involve money (although good rates of payment are believed to be important). The experience of world-class performers, and a principle that Morrison wishes to engender, is that people like to be associated with a successful organisation. In addition, there are initiatives that allow notable 'performers' to be recognised for their efforts.

The future

Perhaps one of the most telling benchmarks that Morrison can point to is the fact that over the last ten years it has been able to simultaneously increase workloads, treble the number of employees, increase its profits and increase the satisfaction levels of customers.

As we would argue, this would not have been possible without having adopted the technique of benchmarking against world-class performers. Whilst using such a technique is not without potential hurdles, the consequence of so doing will be to create the sort of opportunities for improvement that the authors of *Rethinking Construction* envisaged (Construction Industry Task Force, 1998). Not to do so, we would contend, will not only undermine the ability of construction organisations to compete effectively, but will continue the traditional culture in construction whereby valuable effort is needlessly wasted. Most importantly, without benchmarking, people – the so-called 'key resource' – will not be allowed to give their best to the task of providing customers with what they want: a strategic objective for any organisation which wants to become world class.

8.5.7 Case study seven: Working together – a strategy for continuous improvement and benchmarking in John Mowlem plc

Martin Brown, Business Improvement Manager, John Mowlem & Company plc

Introduction

John Mowlem and Company plc is one of the country's leading construction services companies. Founded in 1822 by John Mowlem, the company has an enviable and enduring pedigree second to none. The company has always been innovative in the range of activities undertaken, never more so than the present where it has a wide spectrum of expertise to become a total construction services and solutions provider.

Mowlem is dedicated to the achievement of continuous improvement. This has involved the use of QA (Quality Assurance) using ISO 9000, TQM (total quality management) and, latterly, supply chain management and benchmarking. Fundamental to this maturity and quality development has been use of the Business Excellence Model (EFQM). Described here is an overview of the recent efforts undertaken to ensure that Mowlem is at the forefront of improvement in the British construction industry. As will be explained, in order to achieve this aim, we look to best and better practice from within Mowlem, from the construction industry and from other industries to provide the inspiration and innovation from which lessons can be learnt for introducing best practice.

Continuous improvement

Mowlem has endorsed the principles of SMART Procurement as a basis for continuous improvement. The principles are listed as follows:

- To compete through offering superior underlying value rather than by lower margins
- To establish long-term relationships with key suppliers
- To manage the supply chain during a project through alliances
- To make the provision of value more explicit: design to meet a functional requirement for a through-life cost
- To involve every party in the supply chain in design and cost development
- To encourage a culture of continuous improvement and learning through the supply chain
- To promote greater collaboration by using 'effective' leadership, appropriate training and incentives to employees

Within Mowlem, this will achieved by everybody doing the following:

- Personally managing and continuously improving processes and performance
- Working together in teams, within Mowlem and throughout the supply chain, to deliver excellent service to all customers
- Seeking and adopting best practices
- Measuring and demonstrating improvements in all areas

The Business Excellence Model (EFQM) is used as guidance and as the framework for continuous improvement activities, such as ISO 9000 and the Design Build Foundation.

Moving outside the box

Like the vast majority of organisations that implemented QA using BS 5750 (superseded by ISO 9000; BSI, 1994) we wanted to ensure that quality did not simply end up being seen as a bureaucratic exercise. On the contrary, it should be seen as the beginning of a long-term process of continuous improvement; something that would affect every employee, regardless of task or rank within the organisation. In essence, the development of TQM from QA and the use of the Business Excellence Model was something that we wanted to encourage as being a normal part of the culture that exists.

Benchmarking in Mowlem

Benchmarking is essentially a simple concept. It is about organisations comparing practice and performance on key activities. It involves asking a number of basic questions, including 'where do we need to improve?', 'who is better than us in this area?' and 'why are they better?'. Mowlem has embraced this concept of benchmarking on a number of issues and projects. For example, recent benchmarking activity includes the following:

- Internal benchmarking based upon Business Excellence Model assessments (see later section)
- Benchmarking of internal best practice through a series of forums based around our value management, prime contracting, technical, quality and design activities
- Benchmarking of quality performance across business units within the Northern Division with Quality Management KPIs
- Process benchmarking of our supply chain management strategy with the oil, health, aerospace, IT and construction industries
- Vendor management benchmarking exercise internally and externally with other construction industry organisations
- Contract management benchmarking project to determine value for money in contract management with cross-industry participants led by ITSA, the Government IT Service Agency
- Domestic maintenance customer satisfaction benchmarking, again with leading organisations outside of the construction industry

In addition, we are actively promoting benchmarking within Mowlem and externally within the industry through:

- Promoting the use of the Business Excellence Model in construction, through schemes such as the Design Build Foundation
- The Benchmarking Institute
- The Midlands Construction Forum, whose aim is to encourage business excellence within construction
- Regular participation within quality forums, for example the East Midlands Quality Club
- Mowlem is an active and contributing member of the Best Practice Club, the national benchmarking club

- ACTIVE – a focused initiative aimed at demonstrating that the cost of executing projects for the onshore process and energy industries can be reduced by 30% through improved techniques and the elimination of non-value-adding costs
- Involvement in Construction Best Practice Programme activities, such as the Champions for Change and Best Practice Club's programmes
- Involvement in the Movement for Innovation demonstration projects
- Visits to other organisations and industries as part of the Inside UK Enterprise programme
- Supporting of the Fit for the Future programme to encourage the sharing of best practice across industries

The Benchmarking Institute

A forum to freely stimulate and promote process improvement techniques through:

- Sharing of experience
- Leverage of academic knowledge
- Stimulating discussion
- Trusted networking

The Benchmarking Institute was the result of a post-conference discussion that included several respected individuals closely involved with the process of benchmarking. The opinion of the group was that there was no opportunity for the dedicated benchmarking practitioner to exchange best practice, improving their respective companies or themselves as individuals. Four years on, the Institute now comprises a strong membership of organisations who regularly practise benchmarking; these include companies such as Leatherhead Food RA, Co-operative Bank, John Mowlem & Company plc, ICL, Office of Public Management, BT, The Housing Corporation, Audit Commission, TNT UK Ltd, Royal Mail, Companies House, Rover Group, Conoco and Hyder.

The institute provides a forum for such topics as:

- How business process improvement is undertaken, managed and deployed
- What techniques actually work in organisations
- Academic views and thoughts

- New models developed by the group
- Getting more out of process benchmarking

The main aim of the group is to exchange best practice in a safe environment where none of the developed approaches and deployment plans can be sold on to other parties. There is an underlying philosophy known as the 1–10–100 rule – that is, every hour spent in the forum should give 10 hours of personal benefit that should translate to 100 hours of benefit to the member organisations.

Each year, the Benchmarking Institute makes an award to an institute member in recognition of activity in raising the awareness of benchmarking and business excellence. In 1999, Martin Brown of John Mowlem and Company, was the recipient of this award in recognition of work in raising awareness of benchmarking and the Business Excellence Model within the construction industry; also for continued and proactive participation in quality forums such as the Benchmark Institute, the East Midlands Quality Club, the Midlands Construction Forum and other construction organisations. Previous recipients of the benchmark award include individuals within the Post Office and the NHS.

Midlands Construction Forum

This is a voluntary body which is open to members from all aspects of the construction industry and exists in order to:

- Promote the development of business excellence within the construction industry
- Provide a forum for the free exchange of ideas about the use of improvement tools and techniques
- Provide practical advice and support to fellow members in implementing improvement in their organisations
- Identify, develop and implement best practice across the industry
- Work in support of national bodies that exist to promote business excellence

Working together – applying lessons from benchmarking

The Mowlem supply chain strategy, Working Together, is a systematic approach based upon the best available tools, techniques and practices within Mowlem, within construction and from other industries to achieve significant efficiency improvements and to deliver real solutions.

As the diagram below indicates, there are a number of events (primarily project workshops) that facilitate active participation and learning by all the parties involved. These events (denoted by the thick lines) are explicitly where 'alliances' between these parties occur. What we wish to engender is a climate in which co-operation and sharing is seen to be the way that relationships are conducted; a complete reversal of the adversarialism that many in the industry saw and still see as acceptable.

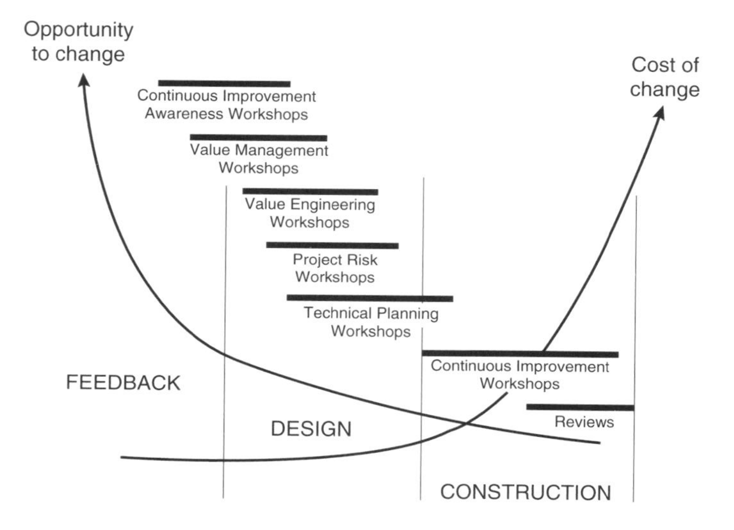

This approach will deliver customer requirements through the rigid application of techniques such as:

- Value management
- Value engineering
- Risk management
- Continuous improvement
- Construction efficiency
- Sustainable construction

Using the Business Excellence Model at project level

Mowlem, having successfully pursued quality in its business activities, reached a level of maturity that needed to be tested, particularly given the current and anticipated changes in the construction industry. A strategic decision was taken to use the Business Excellence Model as the diagnostic tool to:

- Measure progress in pursuit of business objectives
- Continue the maturing of the Mowlem approach to quality issues, moving on from ISO 9000 (BSI, 1994)
- Relate everyday activity to typical behaviours demonstrated by 'excellent' organisations
- Create a methodology and framework for continuous improvement

A strategic partner was sought who:

- Could move beyond the constraints of ISO 9000 (BSI, 1994) and its assessment
- Was an assessor, not an auditor, basing the findings on understanding rather just taking the Model criteria as a standard
- Would add value, directing areas for improvement at business objectives rather than carrying out assessments for assessments sake
- Understood the weaknesses of self-assessment but who didn't want to offer a third-party registration approach

On this basis, BSI Business Solutions was selected, particularly as a true partnership approach was suggested that not only met Mowlem requirements but has proved to be the desired approach throughout the life of the project. An agreed programme of assessments was conducted throughout all business units and central services within the Northern Division of John Mowlem and Co plc against the principal criteria of the Business Excellence Model. The objectives of the programme were to:

- Carry out an assessment against the Business Excellence Model, linking the findings to the objectives and strategies and business unit plans
- Cover the nine criteria of the Business Excellence Model with the focus on key processes and process management
- Transfer Business Excellence Model knowledge and assessment skills to key Mowlem staff

Pilot and plan

A pilot assessment was conducted at one of the regional offices. The results showed that the assessment would work only if key success criteria were followed. From this a detailed schedule for assessments and methodology for the assessments themselves was agreed. The prime issue was to relate all areas of the assessment (i.e. the areas of strength and improvements) against the Divisional and Business Unit objectives and business plans.

Assessments

The assessments which were conducted over a two-day period consisted of one-to-one interviews with managers and facilitated group sessions with staff representatives from all of the departments and projects. It is estimated that over half the staff in the division were involved in this process. The only documentation evidenced was the business strategies and plans, and the framework of management systems (quality, technical, personnel and safety). Assessments concentrated on the activities to achieve the business objectives (Enabler criteria) and the business results achieved (Results criteria). Assessments commenced at director level, moving through the key managers, staff representatives and then back to director level. This approach enabled communication of objectives and strategies and their implementation to be tested up, down and across functions and levels against a set of organisational behaviours that have already been established in various industry sectors.

It was clear that assessors needed a high level of interpersonal skills and business acumen to focus on the business and its objectives, to relate the business to the model rather than the model to the business whilst acting as facilitator. This was even more important given that managers and staff involved ranged from directors to site staff to office-based administration staff such as central credit control and personnel staff. One of the key lessons was that the facilitated approach taken required a level of skills that is not commonly found amongst consultants and/or auditors.

Assessment conclusions

The assessment reports detailed:

- Areas of strength and areas for improvement for each business unit relative to the pursuit of its stated business objectives

- Areas of best practice for sharing across the business and common areas for improvement requiring divisional attention
- Profiling of individual business units using the scoring system to allow internal benchmarking

The project also concluded that the assessments could be targeted not only at the pursuit of business objectives but also at other, external, drivers. In this sense the model can be used to validate activity undertaken to meet other externally imposed requirements.

Continuation strategy

Following the assessments and a review of the lessons learnt, a continuation 'cycle' future strategy for implementation and use of the Model was prepared, as illustrated in the figure below.

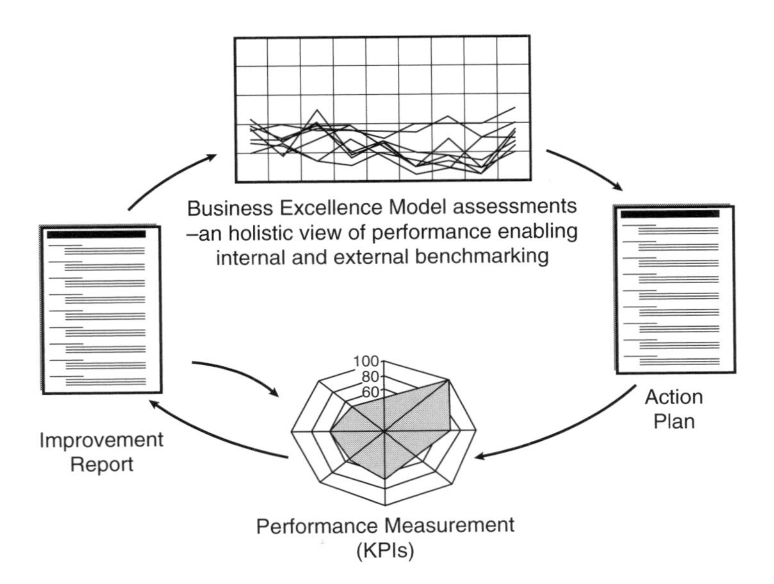

Learning – future use of the Model

Using the Business Excellence Model, against stated business objectives, leads to real, exciting and innovative strategies for future use of the Model within Mowlem, construction and other industry sectors.

As a filter and framework

It is well-acknowledged that the model is a framework for initiatives. Using this proactively, the model can be used as:

(1) An interpreter of external drivers and initiatives
(2) A tool to balance external drivers with business objectives, i.e. to understand the external drivers in the context of business by using the model as a filter (see figure below).

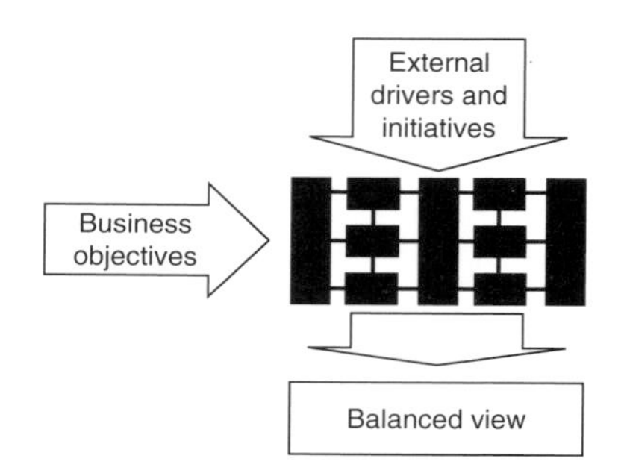

Using at project level

Assessments against business objectives can readily be transferred to project level with project objectives juxtaposed with the higher-level business objectives. The project objectives can be generic and industry-specific, such as prime contracting principles as well as any partnering charter or pact objectives.

As a supplier development tool

By way of example, we are exploring the implementation of a Supplier Excellence Programme, based on the Business Excellence Model, into our supply chain strategy. This approach is well-established within other industry sectors such as aerospace. This is the subject of an exciting research and benchmarking programme.

Conclusion

The experience gained from using the Model as described has enabled both Mowlem and BSI Business Solutions to build upon existing skills and knowledge, and as a direct result, projects are being planned or have already been delivered that:

- Involve a wide cross-section of public and private sectors
- Review improvements to deliver business objectives
- Review and evaluate external drivers such as Best Value, Clinical Governance in the public sector
- Review and evaluate quality and performance standards (ISO, Investors in People, Kings Fund Scheme, Design Build Foundation Registration Scheme) against business objectives
- Focus on internal drivers, which can vary from growing the business to the competence of managers

8.5.8 Case study eight: Amey Supply Chain Services contribution to improvement

Lisa Harris and Chris Sykes, AMEY Supply Chain Services Limited (ASCS)

Introduction

Amey Supply Chain Services (ASCS) is a purchasing agency specialising in the strategic management of non-core spend. It offers a comprehensive range of services to support and develop all elements of the supply chain, including consultancy, outsourcing, transaction management and agency purchasing. Its consultancy service comprises supply chain diagnostic, departmental profiling and benchmarking, strategy formation and change management programmes. Outsourcing involves the transfer of non-core purchasing from the client to ASCS. A contract is then drawn up between the two parties whereby ASCS provides a purchasing service against agreed objectives and measures.

The transaction management service is based on an electronic web-enabled transaction processing engine ('OASIS' – Order and Supply Information System) which is installed and used by clients to bring e-commerce and its many benefits into their supply chain. Agency purchasing is where ASCS uses its considerable purchasing power (based on its large number of clients) to negotiate 'best-in-class' deals on bought-in products and services. These are then offered to an increasing number of clients, and in the spirit of consortia buying, the more clients they have, the better the deals they can negotiate.

Using benchmarking tools

ASCS makes extensive use of both benchmarking and best practice in order to continuously improve its efficiency and effectiveness, and to sustain and improve its position in the market. The word 'benchmarking' describes a range of activities aimed at comparing various aspects of company performance. This might involve comparing prices, costs or processes, and it might take place within the same organisation, between two separate organisations, or amongst a large group of organisations. Furthermore, these organisations might be in similar or completely different industrial sectors. The latter form of benchmarking tends to make use of 'benchmarking clubs' (as in the automotive industry) or specialist consultancy companies.

ASCS is involved in most forms of benchmarking. The fact that the company is highly regionalised, with several sites across the south of England, lends itself to *intra-company benchmarking*. Both the processes and the prices of the separate sites are compared in order to assess performance and to focus efforts at improvement. Price benchmarking is self-explanatory: the prices paid for specific products and services by the separate regions are compared. For each product, the best and worst performing areas can be highlighted and reasons for the differences established. The aim, of course, will always be to achieve the best prices throughout the company, thus improving overall performance. This analysis might also highlight opportunities for leverage: if products or services lend themselves to consolidated purchasing, lower prices can be negotiated.

The second form of intra-company benchmarking, that of processes, is less obvious. In this case, various elements of what are considered to be 'best practice' in purchasing are compared in order to assess the level of competence (or the progress towards achieving best practice status) in each region. Such elements include the level of training of purchasing staff, customer satisfaction, the extent of purchasing involvement in strategic decisions, and whether a continuous improvement approach to management is adopted. The aim here is to transfer learning across the regions and thus improve overall competence.

ASCS also carries out *inter-company benchmarking* (also known as competitive benchmarking) – that is, comparing its performance to other organisations. It does this primarily through the use of a benchmarking consultancy, focusing on the comparison of price. The consultancy offers a computer-based price monitoring service across a wide range of products (approximately 5000), the price data being provided by around 80 subscribing companies. This particular consultancy was chosen because the range of products being monitored most accurately suited ASCS's spend profile. These included computer equipment and peripherals, electrical equipment, office equipment and stationery, office furniture and photocopying services. In essence, the prices paid by ASCS for specific products are compared to the average prices paid by the other subscribers, and to the lowest price paid for any particular product. The price reports produced by the consultants can show ASCS prices against those paid by businesses of a similar type, or against all subscribers.

There are several benefits to using this benchmarking tool. For example, they allow the company to:

- Identify those products that are failing to achieve the most competitive pricing, prompting a renegotiation and putting pressure on suppliers
- Obtain an independent assessment of performance, which is far more credible with customers than one carried out internally
- Use pricing reports to demonstrate to clients areas where the company achieves better-than-average pricing
- Provide an ongoing process for continuous improvement and demonstrate how this will affect the bottom line

Clearly then, benchmarking is an extremely useful tool and an essential ingredient in the process of continuous improvement. There remains, however, a reluctance to adopt this tool in the construction industry. Some of the possible reasons for this are as follows:

- A fear of disclosing competitive information
- A view that the organisation is the best already
- A lack of support and understanding within the organisation
- Difficulty in implementing the process properly because it needs a lot of time and resources. The products benchmarked must obviously be exactly the same specification for the comparison to be of any value

When, as a purchasing organisation, we apply purchasing best practice to the construction and property sectors, we encounter a number of difficulties:

- The 'not invented here' syndrome
- What do purchasing know about construction?
- Construction is different to other service areas

- Employers have a very low expectation of the construction sector, which is not generally perceived as adding value, and therefore seek to drive down the cost of the service
- The project nature of construction makes the relationship too short to develop and deliver partnership benefits to overcome the previous misconceptions

From our own construction experience and discussions with others in the field, the typical client attitude to construction tends to stem from the above. Indeed, a typical construction project may represent 10–20% of the overall project cost, but is not seen as the area which adds value to the whole process. This is in marked contrast to the production equipment inside the structure, which produces the goods that contribute to the company's profitability. So, at best, the structure provides an environment for the value-adding activity to be carried out and at worst it gives rise to snags which have a negative impact on the investment.

Construction, we would suggest, is therefore a service function similar to catering, cleaning, maintenance and business support commodities (stationery, travel, computer equipment, etc.) all of which must be brought together and harnessed to create a profitable organisation.

Key issues

The two key issues, from our own experience are:

(1) Clients need to be more educated to the benefits of a better relationship with their contracting partners. Failure to do so will inevitably result in the continued lack of control of project time-scales and expenditure and long term maintenance and efficiency problems of the finished product.
(2) Construction organisations should see themselves more as service organisations and put more effort into understanding their clients' needs and aspirations. If they create a better and more proactive relationship with clients, they will be seen as 'part of the solution, not part of the problem'. In this way, their involvement will be seen as adding value to the overall process.

Many in the construction industry would probably disagree with our suggestion that they are a service industry. All too often the reaction to such a suggestion is that we are a profession, not a mere service provider. Unfortunately, we are both old enough to remember the banking and insurance sectors adopting a similar attitude. The market today is very different from ten years ago: they have lost the battle and are being replaced by organisations with a clear customer focus. Construction, we would suggest, must move with the times and learn how to do business as part of an increasingly global business community in the twenty-first century and not live in the past. To make this transition, a good appreciation of current business techniques such as benchmarking and best practice must be understood. Both of these techniques have a proven track record in generating significant value-adding benefits (not just cost reductions) to many other business areas as we have previously demonstrated.

The approaches used to benchmarking

In practice we have found that successful benchmarking has four stages. These are:

(1) Your own company (different sites, jobs, locations, etc.)
(2) Your own industry (competitors, etc.)
(3) Industry generally (quality, customer service, defects, client retention, etc.)
(4) World class

Personal experience would suggest that where benchmarking is used in the construction sector, it is generally confined to stages 1 and 2. This is in marked contrast to the service sector, which aspire to operate at stages 3 and 4. Whilst the Government's Construction Best Practice initiative is attempting to break down these barriers at an operational level, many employees find it almost impossible to move from stages 1 and 2 to stages 3 and 4. Why should this be the case? After all, the construction sector pioneered what we call today benchmarking and best practice, with pricing information in publications such as Spons (a national benchmarking tool) and the development of standard forms of contract such as JCT, ICE, RIBA, etc. (industry best practice).

If this is the case, and construction was indeed the market leader in both benchmarking and best practice, why has it now been left behind in the effective use of these business tools? We believe that the answer lies in two areas: education and skills transfer. Construction education and training at all levels through to university and chartered status revolves around design theory and practice and the use and interpretation of standard forms of contract. Generic business skills do not generally form part of the syllabus and 'thinking out of the box' is not something that tends to be encouraged in our experience of recent graduates. Additionally, across the whole supply and services sectors, construction is the least likely to employ personnel from other sectors of industry, thus limiting skills interchange. This, we believe, contributes to the introverted view that prevails within construction and fails to capture the lessons, ideas and best practice from other industries and sectors. Mark Twain once observed 'the human animal is unique for its ability to learn from the mistakes of others and notable for its determined refusal to do so'. The construction sector needs to develop a more enlightened culture where the 'not invented here syndrome' and 'you don't understand construction' attitudes have no place, and good ideas from whatever sector can be taken on board in a positive way to consider how the idea may improve the process or service.

Conclusions

Work by Latham, Egan and, more recently, by the Construction Best Practice Programme, have created an expectation in outside industry that efficiency improvements and cost savings of the order of 30% are achievable in the construction sector. This expectation has driven clients eager to reduce costs to consider this area with renewed enthusiasm. In comparison to other service and commodity areas, the industry's success in the property and construction sector has been much more modest.

Although cost savings of the order of 25% have been achieved, and significant service improvements to the client's organisation, internal professionals often limit access to opportunity areas. More often than not, the professional purchasing organisation's involvement is seen as a threat rather than an opportunity for an independent view. The key to success in this area is to gain senior management buy-in to the process. Failure to negotiate such access will result in wasted time and effort and a failure to deliver the full potential of the piece of work.

Where access to an area is proving difficult for Purchasing, the use of presentations and/or training sessions can inform people of the process which will be used, and hopefully allay fears. The use of such techniques should be sensitive to the internal politics of the organisation, as the object of the exercise is to negotiate supported access, not to ruffle feathers. If all else fails, and very much as a last resort, access must be achieved by senior management mandate. If you find yourself on the receiving end of this type of access, you will quickly understand how General Custer felt at the The Battle of Little BigHorn when surrounded by hostile Indians.

In summary, there are clear benefits to using tools such as benchmarking and best practice. They allow companies to learn from three areas: separate parts of their own organisation, other companies in their industrial sector, and other industries. The key benefit is that they are learning tools, enabling companies to improve the way they operate and become increasingly efficient and effective. In the construction industry, however, their adoption appears to be relatively limited, meaning that this sector is missing out on opportunities to learn from others and continuously improve performance. We would strenuously argue that whatever the reasons for this apparent apathy, the construction sector needs to wake up to the benefits of using these tools. Unless it does so, organisations will discover that others, by using such tools, will be better able to provide the customer with what they want: value at an acceptable cost.

8.5.9 Case study nine: Advancing the management of engineering projects through benchmarking

Ivor Williams and Matthew Seed together with the ECI Benchmarking Steering Committee of the Construction Industry Institute/European Construction Industry Benchmarking Institute

Introduction

Measurement is an essential element of managing projects. All projects have targets for cost, schedule, safety and quality, and most projects are executed according to a project management system and a project execution plan which are intended to ensure that these targets are met. It is straightforward to measure the project outcomes in terms of cost, schedule, safety and quality, but the effectiveness of the systems and procedures in helping to achieve these targets is much less clear. To enable objective measurement of the specific practices which contribute to project outcomes, the systems and procedures can be broken down into these practices, e.g. pre-project planning and change management. The ultimate goal of any benchmarking system must be to identify those practices which influence project outcomes, and to identify those key practices which provide the best return on investment.

The Construction Industry Initiative/European Construction Industry (CII/ECI) Benchmarking Initiative provides, at low cost, a means of objectively measuring the use of those key practices which impact most on project outcome, so providing the industry with performance norms. It also provides a means of benchmarking that practice use against those performance norms and thereby demonstrates the impact and relative value of the key practices. These key practices are referred to as Value Enhancement Practices (VEPs).

The Benchmarking Initiative

The initiative was developed in the USA by the CII and had its first round of data collection in 1996. By early 2000 it had a combined database of approximately 1000 projects, having an average project cost of $50 million and a median cost of $15 million. The ECI joined the initiative in 1997 and by the end of 1999 had 113 projects with an average cost of $104 million and a median cost of $55 million.

The objectives of the initiative are to assist participants to:

- Objectively compare their performance against others
- Show how performance improvements might be achieved
- Quantify the use and value of a given VEP or combinations of VEPs
- Identify norms and trends for the industry and its sectors

Following rigorous research and knowledge-pooling by representatives of leading client and contractor organisations, CII identified the key VEPs which offer the greatest opportunity for performance enhancement on construction projects. These were:

- Team building
- Pre-project planning
- Application of IT to the design/construction process
- Constructability
- Project change management
- Safety

Hence, the initiative benchmarks the effective use of these practices in terms of project performance, i.e. cost, schedule, safety and quality.

The Value Enhancement Practices (VEPs)

The VEPs being benchmarked offer the user the greatest opportunity for performance enhancement. It is recognised, however, that there are other practices which impact on project outcome and hence the list of VEPs needs to expand. The existing VEPs also need to be regularly reviewed to incorporate improvements in the practice generated either by research or by the results of the performance benchmarking.

The current VEPs or best practices as they are sometimes called, have been developed from research and current industry knowledge and are comprehensively documented by CII. ECI has developed VEP summary guidelines which are working tools for implementation of the VEPs. The VEP summaries currently used in the Benchmarking Initiative match the questionnaire, and thereby provide direct guidance to participants on how to improve use of these key practices.

There are two ways in which the VEPs are developed:

(1) New VEPs, which are generated as practices which significantly influence project performance, are identified
(2) Existing VEPs must be updated to reflect industry best practice. (It is particularly relevant that a programme promoting this also practises Continuous Improvement)

The context in which VEPs are developed and used by ECI is shown in the figure below which illustrates the process for introducing new VEPs and the iterative development of the scoring mechanism to ensure accurate correlation between VEP and project performance. This process is currently being used to develop VEPs for risk management, cost estimating, and project controls.

The development of new VEPs is designed to provide a uniform product. The initial briefing for the working group that will consider a new VEP, details the three deliverables expected:

(1) A publication describing current best practice and providing direction in achieving this best practice
(2) A summary document which is a hands-on, project manager's guideline
(3) A questionnaire section to measure and hence benchmark the VEP

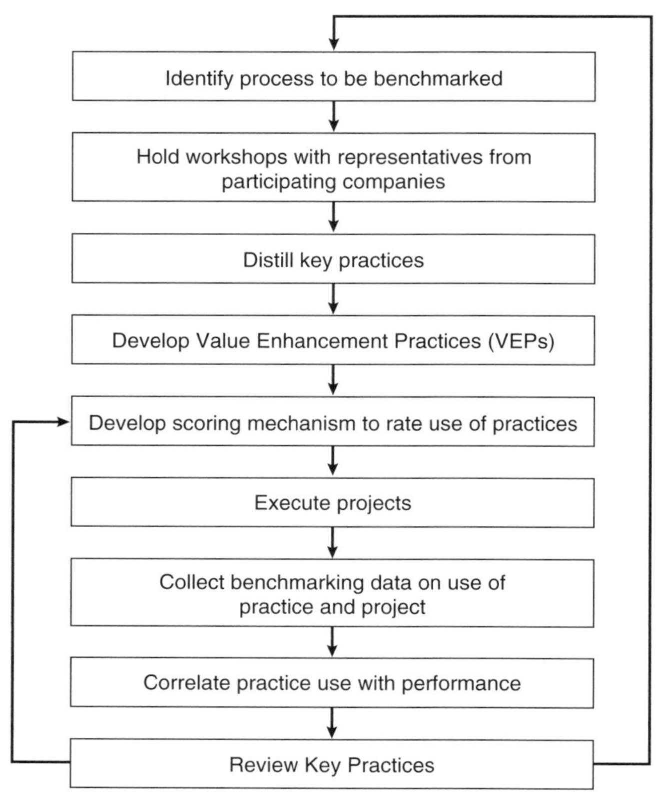

VEP development

The mechanics of benchmarking

Data are gathered using structured questionnaires. Standard data required to calculate project performance measures are collected in the first part of the questionnaire. The key elements of each practice have been extracted from current project management best practice and are addressed by the second half of the questionnaire. In order to further ensure uniformity of data, and maintain confidentiality, companies identify a point of contact through which the questionnaires are distributed and collected. ECI provides training for all these 'Benchmarking Associates' and then also provides an independent audit of submitted data prior to analysis. On completion of the analysis, ECI reformat the output and give full project-by-project interpretation to users.

However, benchmarking should not be seen as an end in itself but should be viewed in a broader sense as part of the continuous improvement cycle. To do so, participating companies need to:

- Consider the benchmarking results in detail.
- Identify with ECI opportunities to improve against the industry norms
- Seize learning opportunities between projects submitted internally, i.e. compare the practice use achieved on each of the projects submitted and look at where lessons can be learned from one's own projects

Since companies executing major engineering projects often operate on a regional or global basis, there is scope for identifying differences in the deployment of VEPs across those regions. Doing this in a systematic and quantifiable way gives enormous opportunities for bringing about improvements.

Results

A large amount of data is generated by the initiative and this is presented to participants in tabular and graphical format. One of the major advantages of the CII/ECI Initiative is the ability to sort data by a number of categories. For example there are four industry categories, namely heavy industrial, infrastructure, light industrial, and buildings. It is possible to make comparisons with the complete data set, or just with the appropriate category. As the size of the database increases, the opportunities for specific comparisons will also increase.

The basic form of results is as metrics, which are measurements derived from questionnaire responses. These metrics are the initial indicators of performance, but it is the more involved analyses, such as Value of Best Practices, which provide clear direction. Brief examples follow which are representative of the overall data set.

Performance metrics

Performance metrics are calculated for cost, schedule, safety and quality. As an example, cost performance is measured in four ways, resulting in four metrics:

(1) Project cost growth
(2) Project budget factor
(3) Phase cost growth
(4) Phase cost factor

Project cost growth measures overall cost growth for any reason and hence is an overall measure of owner and contractor team performance. Project budget factor provides a ratio indicating cost variance for reasons other than approved changes; hence, it provides a good measure of contractor performance. Similar measures are provided for each project phase. An example of these metrics is shown in the figure below. This shows the distribution and averages for the project cost growth metric for ECI owner projects. An illustrative project 'X' shows how the graph can be annotated to indicate project or company performance. Lower y-axis values indicate better performance (negative cost growth is better than positive). The table confirms the plotted data and also shows a 'best in class' data point. (It should be noted that outlier data are excluded from the statistical analysis.)

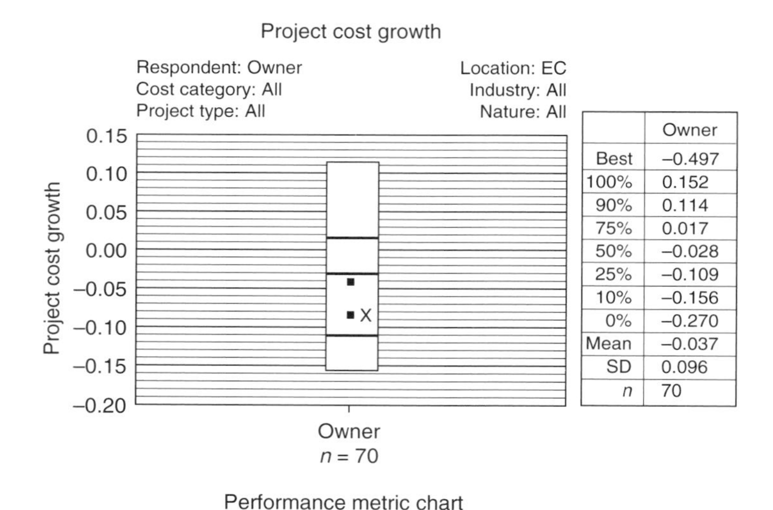

	Owner
Best	–0.497
100%	0.152
90%	0.114
75%	0.017
50%	–0.028
25%	–0.109
10%	–0.156
0%	–0.270
Mean	–0.037
SD	0.096
n	70

Performance metric chart

Practice use metrics

Considerable research into the VEPs has been carried out by CII and indices have been developed to assess, and hence measure, the degree of each practice use. The degree of VEP use is displayed using box plots which show all four quartiles of data, along with the mean result. An example is shown in the following figure, covering constructibility, highlighting the wide variation in use of this VEP.

This shows the distribution and averages for the constructibility practice use metric for ECI owner projects. An illustrative project 'Y'

Constructibility practice use

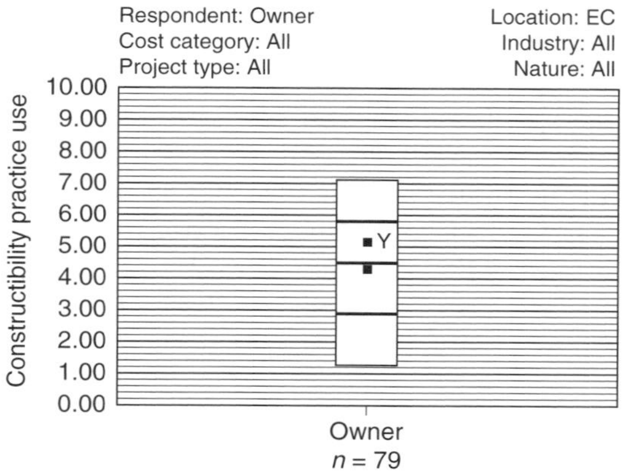

Respondent: Owner
Cost category: All
Project type: All
Location: EC
Industry: All
Nature: All

	Owner
Best	8.67
100%	8.67
90%	7.13
75%	5.75
50%	4.50
25%	2.92
10%	1.25
0%	0.00
Mean	4.28
SD	2.15
n	79

Practice use metric chart

is shown. On practice use metrics, the scale is always 0–10, and a higher score is better.

Value of best practices

When presented with a data set showing that improvement can be pursued in the use of several VEPs, it is helpful to know the value of each practice and the potential for improvements through the increased use of a specific VEP. Hence, measuring the value of VEP use has been an important element of this initiative. The anticipated relationship between practice use and project performance is shown in the figure below. The theoretical plot shows the improvements in project performance as use of the VEP increases; this improvement is reflected not only by a lower mean value, i.e. say lower cost growth, but also by much greater predictability of outcome, demonstrated by the reduced spread of results. The actual results confirm the correlation, although not without some variation, which should disappear as the volume of data captured increases.

An interesting aspect of all the project performance versus practice use correlations has been that there is frequently only very slight variation between the project performance mean result for first quartile or second quartile VEP use, implying diminishing returns as use of the practice approaches high levels. Hence, the decision on using resources to move from second quartile to first quartile use of a specific practice should consider the likely cost benefit.

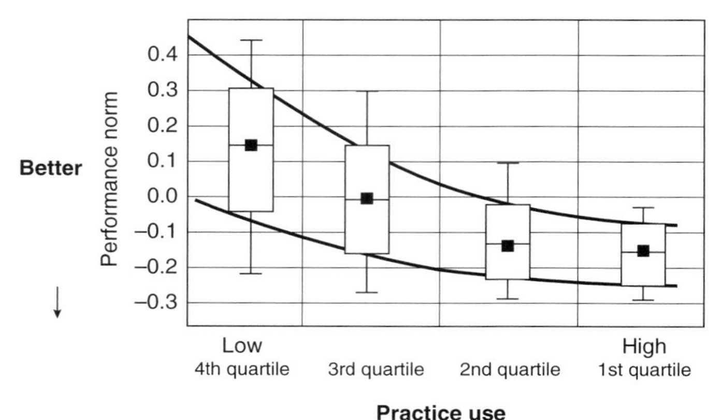

Practice use versus performance norm
This box and whisker plot correlates the project performance statistics against the use of the VEP, categorised by quartiles. Each box portion of the plot represents the middle 50% of the projects and the whiskers extend to the 10th and 90th percentiles, i.e. all but the top and bottom 10% of projects are included for each quartile of practice use.

The above approach has relied on linear correlation of results. However, as benchmarking delivers increased knowledge of the factors that are driving project performance, so the need for more sophisticated analysis of the value of practice use has become evident. The value of VEP use depends on many things, some of which are within the project manager's control, but also some of which are not, for example environmental factors. It is also possible that there is some overlap in the effects due to use of several VEPs.

In order to establish the relative value of the practices, a multiple regression method of statistical analysis is being developed, which will also allow the impact of other external factors to be considered, for example project complexity.

The first set of results generated using a multiple regression method has given an indication of the relative value of the VEPs on project cost growth. Eight separate analyses have been completed; four for owner projects and four for contractor projects. In each case the four analyses break down the projects into the industrial categories detailed earlier.

The data generated by the analyses are presented in pie chart form, a sample of which is shown in the following figure. This chart shows the relative influence of VEPs on project cost growth for owner buildings projects. The VEPs which positively influence project cost growth performance (i.e. reduce cost growth) are, in descending order of importance, pre-project planning, project

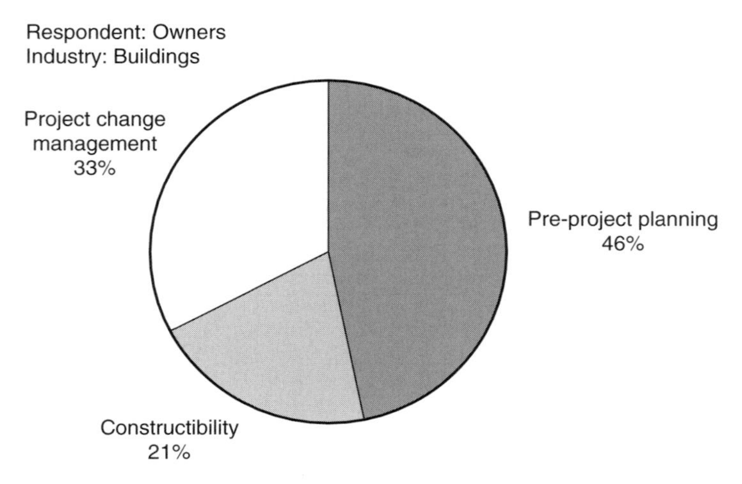

change management and constructibility. This information implies that improvements in practice use should be addressed in this order for the greatest return on investment. It also implies that the remaining VEPs do not have a significant effect on project cost growth for this type of project.

The tables below summarise the results of all the project cost growth analyses, showing the most influential VEPs for each industry group.

Owner projects

Industry	**Key VEPs**
Heavy industrial	Pre-project planning Project change management
Buildings	Pre-project planning Project change management Constructibility
Infrastructure	Project change management Constructibility
Light industrial	Pre-project planning Team building Degree of design completion

Contractor projects

Industry	**Key VEPs**
Heavy industrial	Safety Pre-project planning Design/IT Team building
Buildings	Constructibility Team building
Infrastructure	Pre-project planning Team building
Light industrial	Constructibility Team building

The next stage in developing the Value of Best Practices approach is to assign real values to quantify the relative values derived above. In the case of project cost growth, this would put a figure to the potential savings attributable to increased use of each VEP. Where the potential savings are significant, each company can then decide how to address, and invest in, increased use of each VEP. A similar process can also be applied to the influence of VEPs on performance in schedule, safety and quality.

Trend data

As the initiative matures it becomes possible to determine trends in the data captured; to date, these confirm the trends which would be expected from companies sufficiently progressive to carry out benchmarking of their projects, e.g. reductions in schedule growth and increased VEP use.

Deliverables

The results are presented in three types of report:

(1) Data Reports – for CII projects, presenting aggregated results for projects submitted by CII members, in two categories, USA/Canada and International for ECI Projects, presenting aggregated results for projects submitted by ECI members

(2) Value of Best Practices Reports – presenting summary results for relative influence of VEPs against performance metrics
(3) Key Reports – a customised report confidential to each company, giving specific results for each project and summary results for the aggregated company projects. This is the primary deliverable and provides recommendations on actions to be taken to deliver continuous improvement.

Data Reports

These reports contain all the aggregated data for the complete data set and all current subsets. The total data set includes the performance metrics and practice use metrics data for all 1000 projects. This gives the widest possible comparison, then the smaller data sets such as industry group, project budget or project nature can be used to provide more specific or relevant comparisons. The data reports also include data for each project phase, further adding to the volume of data sets available.

While all of these data sets have potential use, it is likely that each participant will only use a few. In past years the data reports have been presented in hard copy and CD-Rom format. It is now planned to present the Data Reports on an internet web-site, thus allowing participants to select only the relevant data samples for presentation.

Value of Best Practices Report

The presentation used in this report was discussed in an earlier section of this case study. The multiple regression method used involves significant computing time, and this report is produced on an annual basis. To date, analyses have been presented for the relative value of VEPs against project cost growth. The next report will expand to include the influence of the VEPs on schedule, safety and quality.

This report is the key document in identifying where to concentrate effort and resources for maximum return from increased use of VEPs.

Key reports

Each participant company receives a confidential key report at the end of each data round. In the past, data rounds have been annual but, with the advent of internet web-site data collection, this is becoming more frequent. The goal in this area is to provide real-time key reports, where the performance of a project can be immediately returned to the participant.

The components of the key report are:

- Executive summary including recommendations for improved performance
- Individual project performance, metrics and quartiles
- Individual project practice use, metrics and quartiles
- Aggregate performance, metrics and quartiles
- Aggregate practice use, metrics and quartiles

These data can be used by the participant company to assess its performance and priorities for improvement. ECI makes the companies' work much easier by providing an objective analysis of the results and making recommendations for improvement. This analysis also extends to considering trend performance for the participant for as long as there are data available. The final link in the chain is bringing this objective view together with the subjective project experience. This is designed to ensure that the right VEPs are addressed, and that the in-company lessons are learnt from the right projects, ensuring consistent, improved performance.

The benefits of benchmarking

Benchmarking via an external organisation provides an objective mechanism for companies to assess both individual and overall project performance on completed projects with the initial results providing a datum against which continuous performance can be measured on an ongoing basis.

The characteristics of the ECI/CII Benchmarking Initiative which give rise to the benefits to participants are:

(1) Proven mature system
(2) Large database for comparison
(3) Performance measured
  (a) Against broad industry group
  (b) Against peers in sector
  (c) Internally by business unit, contract type, etc.
(4) Interpretation of results highlights
  (a) Areas of excellence
  (b) Improvement opportunities

(5) The VEP development process gives 'best practice' knowledge and guidance to participants
(6) Wide industry participation provides opportunities to share best practice, and for informal networking

The output from the ECI benchmarking exercise provides individual project and overall metrics and rankings indicating both strengths and areas for improvement. The benefits of benchmarking come from a detailed analysis of the output and from converting this into an action plan for improvements. In order to understand all the issues involved, this should be undertaken internally within the organisation or else involve key internal players if conducted externally.

The action plan should set specific targets to be achieved within a given time-scale. The detailed ECI output provides a means of identifying realistic and achievable targets. Although individual performance metrics can only be measured on completion, the processes and practices in place which underpin these can generally be assessed during project execution.

Analysis of individual project outputs can indicate whether practices are being used consistently throughout an organisation. Where this is the case or where areas for improvement have been identified, these are addressed through an internal improvement mechanism. The Value Enhancement Practice guides and other supporting documentation provide a useful starting point in this respect.

The ECI benchmarking exercise also provides a window on new tools and techniques such as the CII Project Definition Rating Index (PDRI) which provides a measure of the quality of underpinning design definition at any point in the design process. The benefits of this type of benchmarking are recognised as being both short term and long term. In the short term, companies benefit by:

- Objective evaluation of strengths and weaknesses
- Framework and guidance for improvement through Value Enhancement Practices
- Increased knowledge base from VEPs and one-to-one exchanges
- Identification of 'gaps' in knowledge and processes

These benefits continue throughout the process, for example when a new VEP is developed. However, there are also some distinct long-term benefits:

(1) Structured mechanism to measure performance enables participants to:
 (a) Identify areas of success and where improvement is needed
 (b) Continuously review best practice
(2) Value of Best Practices report enables participants to prioritise improvement effort and resources
(3) Identification and exploitation of internal 'centres of excellence'
(4) Ability to demonstrate world-class performance.

The consistent approach to data collection requires all participants to have a common understanding of the terminology used and benchmarking acts as a catalyst for this process. Benchmarking is a tool to aid continuous improvement and its importance needs to be understood and supported at all levels within the organisation. With this level of commitment and with an effective implementation strategy, real improvements in project performance can be realised.

Future developments

This benchmarking initiative is seen by participants as having a key role in identifying the areas where effort should be focused to deliver the greatest benefit. The ECI Benchmarking Steering Committee, made up of clients and contractors using the initiative, have moved from the 'why and how to' issues of performing benchmarking, to that of informed practitioners benefiting from the process and determining how to progress the initiative to further enhance project performance.

The developments being pursued either by CII or ECI (or in some cases in parallel) include:

- An effective process for determining and measuring the value of key practices
- Structuring of a web-based questionnaire so that project managers can facilitate continuous data collection and analysis as the project unfolds
- Provision of a supplementary initiative for benchmarking 'small' projects
- Addition of new practice and performance measures; currently, ECI are developing VEPs for risk management, cost estimating

and project controls whilst CII are considering productivity and customer satisfaction.
- Provision of tools and training to assist participants in implementing new/improved value enhancement practices

Conclusions

The benchmarking initiative described is providing, at low cost both in terms of financial stake and resources:

- A measure of project performance against both industry sector and the wider construction industry
- More efficient use of resources through better understanding of the effect of increased use of various value enhancement practices
- Identification of areas for improvement
- Access to the collective experience of major clients and contractors

The initiative is of benefit to developing organisations which have much to learn regarding the use of VEPs, but also to those competing on a global or regional level; the benefit derives from identifying the differences in the deployment of project management resources and tools and hence in project performance. Such differences provide the basis for determining how to improve performance on future projects.

References

Adams, J., Hayes, J. & Hopson, B. (1976) Transitions: Understanding and Managing Personal Change. Martin Robinson, Oxford.

Argyris, C. (1993) *On Organizational Learning*. Blackwell Business, Cambridge, Massachusetts.

Bank, J. (1992) *The Essence of Total Quality Management*. Prentice Hall, Hemel Hempstead.

Ball, M. (1988) *Rebuilding Construction*. Routledge, London.

Banwell Report (1964) *The Placing and Management of Contracts for Building and Civil Engineering Work*. Ministry of Public Buildings and Works, HMSO, London.

Bass, B.M. (1985) Leadership: good, better, best. *Organizational Dynamics*, **13** (winter), 26–40.

Belbin, R.M. (1981) *Management Teams: Why they Succeed or Fail*. Butterworth-Heinemann, Oxford.

Bendell, T., Boutler, L. & Gatford, K. (1997) *The Benchmarking Workout*. FT Pitman Publishing, London.

Blackmon, K., Hanson, P., Voss, C. & Wilson, F. (1999) Being a world organization - what does it mean? In: *Best Practice Process Innovation Management* (ed. M. Zairi), pp. 344–373. Butterworth-Heinemann, Oxford.

Bounds, G., Yorks, L., Adams, M. & Ranney, G. (1994) *Beyond Total Quality Management - Towards the Emerging Paradigm*. McGraw Hill, New York.

BQF (1998a) *Guide to the Business Excellence Model - Defining World Class*. British Quality Foundation, London.

BQF (1998b) *Links to the Business Excellence Model*. British Quality Foundation, London.

Briggs Myers, I. (1987) *Introduction to Type: A Description of the Theory and Application of the Myers Briggs Type Indicator*. Consulting Psychologists Press, Palo Alto, California.

Brown, T. (1993) *Understanding BS 5750 and Other Quality Systems*. Gower Books, Aldershot.

BSI (1994) Series of Standards for the Management of Quality. BS EN ISO 9000. British Standards Institution, London.

BSI (1995) *Quality Management and Quality Assurance - Vocabulary*. BS EN ISO 8402 (formerly BS 4778: Part 1, 1987/ISO 8402, 1986). British Standards Institution, London.

BSI (1996) *Environmental Management Systems (Specification with Guidance for Use)*. BS EN ISO 14001. British Standards Institution, London.

Burnes, B. (1996) *Managing Change: a Strategic Approach to Organisational Dynamics*. Pitman Publishing, London.
Buzzel, R.D. & Gale, B.T. (1987) *The PIMS Principles: Linking Strategy to Performance*. The Free Press, New York.
Camp, R.C. (1989) *Benchmarking: The Search for Industry Best Practices that Lead to Superior Performance*. ASQC Quality Press, Milwaukee, WI.
Cargill, J. (1994a) Orient express. *Building Magazine*, 2nd December, 33.
Cargill, J. (1994b) Macs packs. *Building Magazine*, 29th July, 34.
Carlzon, J. (1987) *Moments of Truth*. Ballinger Publications, Cambridge, Massachusetts.
Chartered Institute of Building (1995) *Time for real improvement: learning from best practice in Japanese construction R&D*. Report of the DTI Overseas Science and Technology Expert Mission to Japan, December 1994. CIOB Publications, Ascot.
Christopher, M., Payne, A. & Ballantyne, D. (1991) *Relationship Marketing – Bringing Quality, Customer Service and Marketing Together*. Butterworth-Heinemann, Oxford.
Christopher, M. & Yallop, R. (1991) *Audit your Customer Service Quality*. Cranfield School of Management, Bedford.
CIRIA (1998) *Benchmarking for construction, a strategic view*. Project report 69, Construction Industry Research and Information Association, London.
Codling, S. (1992) *Best Practice Benchmarking – a Management Guide*. Gower Publishing, Aldershot, Hampshire.
Construction Best Practice Programme (CBPP) literature, Watford.
Construction Industry Task Force (1998) *Rethinking Construction*. Department of the Environment, Transport and the Regions, London.
Crainer, S. (1996) *Key Management Ideas: Thinking that Changed the Management World*. Pitman Publishing, London.
Cross, R. & Leonard, P. (1994) Benchmarking: a strategic and tactical perspective. In: *Managing Quality* (ed. B.G. Dale), pp. 497–513. Prentice Hall, Hemel Hempstead.
Dale, B.G. (1994) Japanese total quality control. In: *Managing Quality* (ed. B.G. Dale), pp. 80–116. Prentice Hall, Hemel Hempstead.
Dale, B.G. & Cooper, C. (1992) *Total Quality and Human Resources: an Executive Guide*. Blackwell Business, Oxford.
Dale, B.G. & Boaden, R.J. (1994) The use of teams in quality improvement. In: *Managing Quality* (ed. B.G. Dale), pp. 514–30. Prentice Hall, Hemel Hempstead.
Dale, B.G., Boaden, R.J. & Lascelles, D.M. (1994) Total quality management: an overview. In: *Managing Quality* (ed. B.G. Dale), pp. 3–40. Prentice Hall, Hemel Hempstead.
DETR (2000) *KPI Report for The Minister for Construction*. Department of the Environment, Transport and the Regions, London.
Economic Intelligence Unit (1992) *Making Quality Work: Lessons from Europe's Leading Companies*. Economic Intelligence Unit, London.

EFQM (1999) *Assessing for Excellence*. European Foundation for Quality Management, Brussels.

Ferry, J. (1993) *The British Renaissance: Learn the Lessons of How Six British Companies are Conquering the World*. William Heinemann Limited, London.

GAC (1992) *Management Practices – US Companies Improve Performance Through Quality Efforts*. General Accounting Office/NSIAD-91-190.

Hackman, J.R. & Oldman, G.R. (1980) *Work Redesign*. Addison-Wesley Publishing Inc., Reading, Massachusetts.

Hotten, R. (1998) *Formula 1, the Business of Winning*. Orion Business Books, London.

Jaques, E. (1952) *The Changing Culture of a Factory*. Drydren Press, New York.

Judd, V.C. (1987) Differentiate with the 5th P: People. *Industrial Marketing Management*, **16**, 241–7.

Juran, J.M. (1993) Made in USA: a renaissance in quality. *The Harvard Business Review*, **71**, No. 4, 42–50.

Karlof, B. & Ostblom, S. (1993) *Benchmarking: a Signpost to Excellence in Quality and Productivity*. John Wiley and Sons, Chichester.

Kearns, D.T. & Nadler, D.A. (1992) *Prophets in the Dark: How Xerox Reinvented Itself and Beat the Japanese*. Harper, New York.

Kennedy, C. (1994) *Managing with the Gurus*. Century Books, London.

Kormanski, C. & Mozenter, A. (1987) A new model for team building: a technology for today and tomorrow. In: *The 1985 Annual Developing Human Resources Conference*. University Associates, San Diego, California.

Kotter, J. & Heskett, J. (1992) *Corporate Culture and Performance*. Free Press, New York.

Lascelles, D.M. & Dale, B.G. (1993) *The Road to Quality*. IFS Limited, Bedford.

Latham, M. (1994) *Constructing the Team*. The Stationery Office, London.

Levitt, T. (1983) *The Marketing Imagination*. The Free Press, New York.

McCabe, S. (1997) Using suitable tools for researching what quality managers in construction organisations actually do. In: *Journal of Construction Procurement*, **3**, No. 2, 72–87.

McCabe, S. (1998) *Quality Improvement Techniques in Construction*. Addison-Wesley-Longman, Harlow.

McCabe, S. (1999) *Managing quality in construction – an interpretive study*. PhD thesis, University of Birmingham.

McCabe, S. & Robertson, H. (2000) Striving to be the best: a case study of construction benchmarking world-class organisations. In: *Proceedings of the 16th Annual ARCOM Conference*, Glasgow Caledonian University, pp. 355–64.

McGeorge, D. & Palmer, A. (1997) *Construction Management – New Directions*. Blackwell Science, Oxford.

Macdonald, A.M. (1972) *Chambers Twentieth Century Dictionary*. W R Chambers, London.

Manson, M.M. & Dale, B.G. (1989) The operating characteristics of quality circles and yield improvement teams: a case-study comparison. *European Management Journal*, **7**, 370–96.

Mayo, A. (1993) Learning at all organizational levels. In: *Learning More about Learning Organisations* (ed. P. Sadler). AMED Publications, London.

Morrison, S.J. (1994) Managing quality: an historical review. In: *Managing Quality*, (ed. B.G. Dale), pp. 41–79. Prentice Hall, Hemel Hempstead.

Oakland, J.S. (1993) *Total Quality Management, the Route to Improving Performance*. Butterworth-Heinemann, Oxford.

Oakland, J.S. (1999) *Total Organizational Excellence, Achieving World-Class Performance*. Butterworth-Heinemann, Oxford.

Parasuraman, A., Zeithaml, V.A. & Berry, L.L. (1988) SERVQUAL: A multiple-item scale for measuring consumer perceptions of service quality. *Journal of Retailing*, **64**, No. 1 (spring), 12–40.

Pedler, M., Burgoyne, J. & Boydell, T. (1991) *The Learning Company: a Strategy for Sustainable Development*. McGraw Hill, London.

Peters, G. (1994) *Benchmarking Customer Satisfaction*. Financial Times/Pitman Publishing, London.

Peters, T. & Austin, N. (1985) *A Passion for Excellence*. Collins, London.

Peters, T. & Waterman, R. (1982) *In Search of Excellence: Lessons from America's Best-run Companies*. Harper & Row, New York.

Pickrell, S., Garnett, N. & Baldwin, J. (1997) *Measuring Up, a Practical Guide to Benchmarking in Construction*. Construction Research Communications Ltd, Watford.

Porter, L. & Tanner, S. (1996) *Assessing Business Excellence*. Butterworth-Heinemann, Oxford.

Porter, M. (1980) *Competitive Strategy: Techniques for Analysing Industries and Competitors*. The Free Press, New York.

Porter, M. (1985) *Competitive Advantage*. The Free Press, New York.

Pursell, C. (1994) *White Hart – People and Technology*. BBC Books, London.

Roethlisberger, F.J. & Dickson, W.J. (1939) *Management and the Worker*. Harvard University Press. Cambridge, Massachusetts.

RSA (1995) *Tomorrow's Company*. Royal Society for the Encouragement of Arts, Manufacture and Commerce, London.

Sadler, P. (1995) *Managing Change*. Kogan Page, London.

Sako, M. (1993) *Prices, Quality and Trust*. Cambridge University Press, Cambridge.

Scherkenbach, W.W. (1986) *The Deming Route to Quality and Productivity*. Mercury, London.

Schonberger, R.J. (1990) *Building a Chain of Customers*. Free Press, New York.

Schutz, W. (1978) *FIRO Awareness Scales Manual*. Consulting Pychologists Press, Palo Alto, California.

Seddon, J. (1997) *In Pursuit of Quality*. Oak Tree, Dublin.

Senge, P., Kleiner, A., Roberts, C., Ross, R.B. & Smith, B.J. (1994) *The Fifth Discipline Fieldbook: Strategies and Tools for Building a Learning Organization*. Nicholas Brealey, London.

Simon Report (1944) *Report on the Management and Placing of Building Contracts*. Ministry of Works, HMSO, London.

Sims, D., Fineman, S. & Gabriel, Y. (1993) *Organizing and Organizations, an Introduction*. Sage Publications, London.

Sirkin, H.L. (1993) The employee empowerment scam. *Industry Week*, 18 October, 58.

Stahl, M.J. (1995) *Management – Total Quality in a Global Environment*. Blackwell Publishers, Cambridge, Massachusetts.

Taylor, A. (1996) Japanese go for the inglenook look. *Financial Times*, 1 March, 5.

Taylor, F.W. (1911) *The Principles of Scientific Management*. Harper, New York.

Teare, R., Atkinson, C. & Westwood, C. (1994) *Achieving Quality Performance, Lessons from British Industry*. Cassel, London.

Tuckman, B.W. & Jensen, M.A. (1977) Stages of small group development revisited. *Group and Organizational Studies*, **2**, 419–27.

Walton, M. (1989) *The Deming Management Method*. Mercury Business Books, London.

Williams, R. & Bersch, B. (1989) *Proceedings of the First European Quality Management Forum*, pp. 163–72. European Foundation for Quality Management, Brussels.

Womack, J.P. & Jones, D.T. (1996) *Lean Thinking – Banish Waste and Create Wealth in your Corporation*. Simon and Schuster, New York.

Wood Report (1975) *The Public Client and the Construction Industry*. HMSO, London.

Zairi, M. (1996) *Benchmarking for Best Practice – Continuous Learning through Sustainable Innovation*. Butterworth-Heinemann, Oxford.

Notes

1. It is acknowledged that, until now, the sport at this level is one where there are no women competitors.
2. Hotten makes the point that Formula One, is effectively, a showcase for the major mass producers of cars, an industry, in which, he contends, being acknowledged as world class is essential to continued commercial success.
3. Whilst this may seem to be a case of 'coals to Newcastle', it is not without precedent. As has been described earlier in this chapter. Nissan cars made in Sunderland are now exported to Japan.
4. Whilst this report was written by the Construction Task Force, Sir John Egan, because he was Chair of this group, has become the focus of attention. As a consequence, it has become commonplace to refer to this report as the 'Egan Report'. This is a convention which is continued in this book.
5. This is the person widely credited with instructing Japanese industry post World War II in how to become excellent. What his teachings were, and how they were applied, will be dealt with in detail in Chapter 3.
6. CALIBRE is a system that provides a way for the mapping of on-site processes, and by providing the information recorded to BRE, to be provided with 'real-time feedback to site managers to help them to remove barriers to productivity, eliminate waste and improve value-adding activities' (Construction Task Force, 1998: p. 13).
7. Value management can, according to the Egan Report, reduce costs by up to 10%, and is defined as 'a structured method of eliminating waste from the brief and from the design before binding commitments are made' (Construction Task Force, 1998: p. 13).
8. This is one of the many definitions that exist for the word benchmarking. Even though this book will refer to others, this one summarises the main objective of its use.
9. The details that appear are provided with the kind permission of the Construction Best Practice Programme. Whilst every effort has been made to ensure their accuracy, the author takes no responsibility for any changes that may be made by the programme subsequent to publication.
10. The CIB was established as a direct response to Sir Michael Latham's report *Constructing the Team*.
11. The Construction Best Practice Programme does not preclude any

initiative but in its literature includes the following 'major business-improvement areas or themes': benchmarking; briefing the team; choice of procurement route; culture and people; health and safety; information and technology; integrating design and construction; lean construction; partnering and team development; risk management; standardisation and pre-assembly; supply chain management; sustainable construction; value management; whole-life costing.

12 The rationale for commissioning the Construction Industry Task Force to write *Rethinking Construction*.

13 This report provides an invaluable source of information and advice on what KPIs are and, more especially, the methodology for using them.

14 'Commit to invest' is the point at which the client decides to embark upon the project by giving authorisation to the design team to proceed with conceptual drawings.

15 'Commit to construct' is the point at which the client gives authorisation to the project team to allow commencement of construction on site.

16 Elton Mayo, because he led the research team at the Hawthorne factory, is the person credited with having discovered the importance of what is usually called the 'Human relations movement'.

17 The use of the word 'ends' is metaphorical. *Kaizen* is about continuous improvement; it must never end.

18 This will be more fully defined in the next chapter, but is usually considered to be the way(s) that people accept that tasks are carried out on a day-to-day basis.

19 Quality manager – a title that will be used to describe change agents – is commonly used in all industries. It is, however, not uncommon for these managers to also be referred to as 'Business Improvement Managers', or notably, as exists in two different organisations, 'The Change Process Engineer', and 'The Cultural Development Facilitator'.

20 This is point number 12 of the 14 principles of management that Deming recommended (see McCabe, 1998: p. 34).

21 Note that the EFQM Business Excellence Model (see Chapter 7) includes two criteria sections that concern 'people' and which, notably, provide 18% of the overall score that an organisation can attain.

22 As explained already, the philosophy of TQM, whilst being about the 'end' customer, uses the concept of 'internal customers' to ensure that at every stage in the *total* process, there is an obsession with ensuring that what you provide to them is exactly what they want or expect.

23 Hackman and Oldman stress that it is important to be aware of the individual aspirations of workers. As a consequence, they have provided what they call the Job Diagnostic Survey (JDS) which is a measure (on a scale of 1–7) of their job characteristics in order to arrive at an overall measure of job enrichment, the Motivating Potential Score (MPS):

$$MPS = \frac{(\text{skill variety} + \text{task identity} + \text{task significance})}{3} \times \text{autonomy} \times \text{feedback}$$

24 This was one of the most crucial findings that emerged from the so-called 'Hawthorne experiments' (see Roethlisberger & Dickson, 1939).
25 This was the first company outside Japan to win the Deming prize in 1989.
26 This is often a task that the quality manager may have to perform.
27 Belbin has already been described. However, a number of alternative methods exist such as the *Myers Briggs Type Indicator* (Briggs Myers, 1987), *FIRO-B* (Fundamental Interpersonal Relations Orientation-Behaviour) (see Schutz, 1978), *The inclusion, control and openness cycle* and *The five 'A' stages of teamwork* (Oakland, 1999)
28 Initiatives such as *The Learning Society* have become an integral part of British Government policy since 1997.
29 Attention is drawn to this distinction in recognition of the fact that whilst customers normally pay for what they receive, this may not always be the case. For instance, the provision of health through the NHS is not a service that all those who avail of it must pay for through indirect taxation (i.e. the young and unemployed). However, every person who avails of the NHS is regarded as being an equal customer; in effect, society is the customer.
30 In the past, this argument was believed to apply only to organisations operating in the private sector. Public sector organisations are now expected to ensure that they are able to prove their ability to satisfy those 'customers' who receive the products or services they provide. In the case of organisations which fail in their ability to satisfy clients (or show effort to provide value), there are, increasingly, mechanisms for dealing with this, such as reduction in funding or loss of authority to continue operating.
31 As described in Chapter 4, the culture of the organisation will have a large influence on the relationship that senior managers have with those employees carrying out day-to-day tasks. This, therefore, will have a bearing upon how much knowledge the senior managers possess with respect to what actually occurs.
32 Primary are what can be regarded as 'main' processes which cover a major activity such as the way that materials are used in a production process. However, in order to get these materials from the supplier on time, to the right specification and to the correct location, involves a number of secondary processes (which can, if necessary, be broken down again).
33 Procedures are a vital part of a quality system (see McCabe, 1998 for a full explanation of how quality systems should be used).
34 The author was told a story of how, in a car manufacturing plant, forms recording information which no one wanted, kept being sent to a particular office. As soon as these forms arrived they were disposed of. No one bothered to ask why this was occurring until the person

who related the story started work and asked precisely this question. As he discovered, the form had been used to record chassis numbers which were required in the post *World War II* steel shortages; no one had told those involved that the need to record this information had been dispensed with many decades before. Even more farcically, the employee stated that those who were still recording the information had a cabinet full of forms in case they ran short. Even more amusingly, because the need to see the number required climbing into a darkened pit, there was another cabinet full of torches and spare batteries.

35 Usually on the basis of minimum qualifications and experience to indicate attainment of certain skills.

36 Even though the use of procedures may be considered counter-productive in some tasks, especially those which involve the use of high levels of expertise (such as, for instance, brain surgery), the use of documented expertise will assist those who attempt to carry out similar functions.

37 Consensus is believed to be very important; best practice is not something that should be imposed.

38 The author witnessed the use of this method by an automotive supplier that successfully employed people with learning difficulties to assemble components for a model which was acknowledged to be of an 'extremely high standard of quality'.

39 It is usually the case that the larger the organisation, the larger its marketing and sales department, and as a result, the size of the budgets that are necessary.

40 The form that the measures used take.

41 One of the greatest criticisms that is often levelled at public sector organisations is, because of standard rules and procedures - often called *standing orders*, the inability of officers (a bureaucratic term that is used to describe those who act on behalf of such bodies) to alter what they or others do for their clients, even when it results in dissatisfaction.

42 This may be due to price considerations, dissatisfaction about the service they currently receive, or simply the belief that a change is worth trying.

43 There are ethics that must be considered in doing this, most particularly with respect to asking sensitive questions about competitors. It may be useful for a third-party specialist organisation to do this.

44 This could be an identified critical success factor.

45 Examples of these companies are Borden, Campbell, Coca-Cola, Kraft General Foods, Nabisco, Proctor and Gamble, Safeway.

46 Consider the example of, for instance, out-of-season strawberries from Israel, where in order to create maximum return, the supplier chain uses high-cost forms of transit to ensure that the produce is on the shelf within the shortest time possible.

47 For instance, Federal Express, Florida Power and Light (first non-Japanese Deming Prize winners), 3M, BASF, British Airways, Fuji Xerox, Komatsu, Matsushita Electrical and Toyota.
48 The examples of the development of microelectronic products such as camcorders and miniaturised audio equipment being the most widely recognised developments of this obsession.
49 In recent years in the UK, the person most readily identified in this way is Richard Branson.
50 This, of course, represents the main 'critical success factor'.
51 At the time of writing, Marks and Spencer and Sainsburys - because many of their traditional customers are shopping elsewhere - are experiencing unanticipated reductions in profit.
52 An influential writer on the reasons why industries and countries enjoy competitive advantage; see, for instance, *Competitive Strategy* (Porter, 1980) and *Competitive Advantage* (Porter, 1985).
53 Even though the expression 'world class' really applies to level six, regional quality awards now exist by which *any organisation* - regardless of its purpose - can be assessed to demonstrate its ability to achieve excellence (these are described in detail in a subsequent section).
54 This involves managers relinquishing their traditional 'command and control' role in order to allow a much greater degree of empowerment.
55 The author learnt that when Nissan were developing the Micra, their buyers bought versions of every model that the Micra would be competing against. The reason for this was that these models were shipped back to Japan so that every detail of them (no matter how minute) could be analysed in detail. In particular, Nissan wanted to learn the defects on the models they examined so as to eliminate them in the Micra. The final product was, as various awards testify, widely acclaimed to represent excellence at a price which was extremely competitive when compared to existing models.
56 It is a little-known fact that Deming, whilst being honoured, disagreed with the creation of this prize. As he believed, the fact that someone wins means that others must lose. This, he felt is divisive, and more importantly, because winning appears to be the pinnacle of success, serves to undermine the fundamental principle of continuous improvement.
57 JUSE provides a definition of CWQC as being 'a system of activities to ensure that quality products and services required by customers are economically designed, produced and supplied while respecting the principle of customer-orientation and the overall public well-being. These quality assurance activities involve market research, research and development, design, purchasing, production, inspection and sales, as well as other related activities inside and outside the company. Through everyone in the company understanding both the statistical concepts and methods, through their application to all the

aspects of quality assurance and through the repeating cycle of rational planning, implementation, evaluation and action, CWQC aims to accomplish business objectives.' (Quoted in Porter & Tanner, 1996: p. 34)

58 I have deliberately replaced the word 'company' with 'organisational' in order to stress that non-profit-making bodies are encouraged to apply to be assessed for this award.

59 This is a point which was reinforced in Chapter 4.

60 BQF (The British Quality Foundation) is the organisation which propagates the use of the EFQM Excellence Model in the UK (see section 7.6).

61 The reader's attention is drawn to the fact that an organisation may be one that depends upon the voluntary contribution of people who may not draw a wage or salary.

62 The BQF was formed in November 1992 in order to achieve in the UK what EFQM was carrying out in Europe – that is the promotion and use of organisational excellence as a means to achieve superior performance through continuous improvement.

63 This information is based upon discussions with assessors of the Business Excellence Model.

64 Site visits usually follow the submission during which assessors have an opportunity to speak to employees and see actual methods of management in practice.

65 This was an initiative that developed from a report sponsored by the Royal Society for the Encouragement of Arts, Manufacture and Commerce in 1993. This report, carried out by a team of people representing 24 of the UK's best performing companies, sought to examine how business leaders could be assisted in improving the competitive performance of UK organisations. As a direct consequence of the report *Tomorrow's Company* (RSA, 1995), the Centre for Tomorrow's Company was created in June 1996 with the specific purpose of encouraging leaders of organisations to achieve 'sustainable success and world class performance' and can be contacted on 020 7930 5150. (BQF, 1998a: p. 17).

66 The word 'corporate' is used here in the widest sense and is intended to include non-profit-making organisations which, for instance, may be based in the public sector.

67 Investors in People is a standard, which since 1990, has existed in order to encourage organisational improvement through the development of employees. This standard is delivered by the national network of Training and Enterprise Councils (TECs) in England and Wales, Local Enterprise Companies (LECs) in Scotland and the Training and Employment Agency (T&EA) in Northern Ireland. Specifically, the Investors in People Standard seeks to assist organisations in achieving the following four principles (© *Investors in People*, 1998):

(1) An Investor in People makes a commitment from the top to develop all employees to achieve its business objectives
(2) An Investor in People regularly reviews the needs and plans, the training and development of all employees
(3) An Investor in People takes action to train and develop individuals on recruitment and throughout their employment
(4) An Investor in People evaluates the investment in training and development to assess achievement and future effectiveness

68 Management Standards – Management Charter Initiative and Vocational Qualifications were originally introduced in 1989 following research into the work of 6000 managers. Having been revised, they were reissued in 1997.
69 The Charter Mark was launched in 1991 as a part of the Government's Citizen's Charter programme. The objective of this programme was that, in order to win this award, organisations – especially those that serve the public – must demonstrate consistent ability to provide extremely high levels of service.
70 The rationale for the scores that these parts have been allocated is based upon the judgement and experience of those who have analysed the use of other models.
71 Note: copyright rules do not allow more detailed description than appears here.
72 A 50% score for approach would require that there is evidence that the enabler is systematically applied, regularly reviewed and allows the organisation to meet its objectives.
73 A 25% score for deployment would be appropriate if the enabler is only being used to a quarter of its potential.
74 The author's experience has shown that the more often one carries out this process, the more confident one becomes in interpreting the accuracy of scores.
75 A score such as this would be appropriate for a result if there was evidence that it covered at least three years and that it showed a consistently good performance against both internal and external benchmarks.
76 A score such as this would be given to a sub-criterion, the result of which covers only *some* (a word that EFQM uses) of its activities and relevant areas.
77 Now commonplace as part of the assessment process for those organisations which apply to be considered for an excellence award such as EFQM.
78 This task can be a valuable source of organisational learning to discover how processes are really carried out.
79 It is important that as wide a range of people as possible should be involved.
80 The amount of time and resources will depend upon the size of the organisation and expertise of those involved.

Index